T0201433

Pump Wisdom

Pump Wisdom

Essential Centrifugal Pump Knowledge for Operators and Specialists

Second Edition

Robert X. Perez
San Antonio, Texas
USA

Heinz P. Bloch
Montgomery, Texas
USA

The Global Home of Chemical Engineers

Copyright © 2022 by the American Institute of Chemical Engineers, Inc. All rights reserved.

A Joint Publication of the American Institute of Chemical Engineers and John Wiley & Sons, Inc.

Edition History
1st edition: Wiley 2011

Published by John Wiley & Sons, Inc., Hoboken, New Jersey.

Published simultaneously in Canada.

No part of this publication may be reproduced, stored in a retrieval system, or transmitted in any form or by any means, electronic, mechanical, photocopying, recording, scanning, or otherwise, except as permitted under Section 107 or 108 of the 1976 United States Copyright Act, without either the prior written permission of the Publisher, or authorization through payment of the appropriate per-copy fee to the Copyright Clearance Center, Inc., 222 Rosewood Drive, Danvers, MA 01923, (978) 750-8400, fax (978) 750-4470, or on the web at www.copyright.com. Requests to the Publisher for permission should be addressed to the Permissions Department, John Wiley & Sons, Inc., 111 River Street, Hoboken, NJ 07030, (201) 748-6011, fax (201) 748-6008, or online at http://www.wiley.com/go/permission.

Limit of Liability/Disclaimer of Warranty: While the publisher and author have used their best efforts in preparing this book, they make no representations or warranties with respect to the accuracy or completeness of the contents of this book and specifically disclaim any implied warranties of merchantability or fitness for a particular purpose. No warranty may be created or extended by sales representatives or written sales materials. The advice and strategies contained herein may not be suitable for your situation. You should consult with a professional where appropriate. Further, readers should be aware that websites listed in this work may have changed or disappeared between when this work was written and when it is read. Neither the publisher nor authors shall be liable for any loss of profit or any other commercial damages, including but not limited to special, incidental, consequential, or other damages.

For general information on our other products and services or for technical support, please contact our Customer Care Department within the United States at (800) 762-2974, outside the United States at (317) 572-3993 or fax (317) 572-4002.

Wiley also publishes its books in a variety of electronic formats. Some content that appears in print may not be available in electronic formats. For more information about Wiley products, visit our web site at www.wiley.com.

Library of Congress Cataloging-in-Publication Data

Names: Perez, Robert X., author. | Bloch, Heinz P., 1933– author.
Title: Pump wisdom : essential centrifugal pump knowledge for operators and
 specialists / Robert X. Perez, San Antonio, Texas, USA , Heinz P. Bloch,
 Montgomery, Texas USA.
Description: Second edition. | Hoboken, NJ., USA : Wiley, [2022] | Includes
 index.
Identifiers: LCCN 2021046316 (print) | LCCN 2021046317 (ebook) | ISBN
 9781119748182 (hardback) | ISBN 9781119748199 (adobe pdf) | ISBN
 9781119748236 (epub)
Subjects: LCSH: Pumping machinery.
Classification: LCC TJ900 .B6483 2022 (print) | LCC TJ900 (ebook) | DDC
 621.6/9–dc23
LC record available at https://lccn.loc.gov/2021046316
LC ebook record available at https://lccn.loc.gov/2021046317

Cover Design: Wiley
Cover Images: © Matveev Aleksandr/Shutterstock; engineer story/Shutterstock;
prabhjits/Getty Images; Grassetto/Getty Images

Set in 9.5/12.5pt STIXTwoText by Straive, Pondicherry, India

SKY10033264_021122

"Dedicated to the reliability technicians and engineers who will not rest until enhanced process pump safety and reliability become the new normal."

Contents

Preface

In the 10 years since the first edition of this book was compiled, upgrading in accordance with this text was largely practiced by a few best-in-class performers ("BiCs"). The feedback from these fluid machinery users was very gratifying and the publishers expressed interest in a second, expanded Edition. Robert Perez was asked to join the two-man writing team and the results are found in this updated Second Edition text. True to the subtitle we selected, you will find this volume to contain Essential Centrifugal Pump Knowledge for Operators and Specialists.

Next to electric motors, centrifugal pumps are the most frequent rotating machine man has built and put to use. Centrifugal pump users have access to hundreds of books and many thousands of articles dealing with pump subjects. So, one might ask, why do we need *this* text? We co-authors are certain that we need this text because an unacceptably large number of centrifugal process pumps fail catastrophically every year. An estimated 95% of these are repeat failures and most of them are quite costly, or dangerous, or both.

Essentials concentrate on explaining the many elusive failure causes that manifest themselves as repeat failures. We saw it as our mandate and task to clearly map out permanent remedial action. Our intent was to steer clear of the usual consultant-conceived generalities and give you tangible, factual, and well-defined information throughout.

As any close review of what has been offered in the past will uncover, many texts were written to primarily benefit one particular job function, ranging from centrifugal pump operators to pump designers. Some books contain a hidden bias, or they appeal to a very narrow spectrum of readers; others are perhaps influenced by a particular centrifugal pump manufacturer's agenda. Give *this* text a chance, you will see that it is different. You will not find it in other texts. We gave it the title *Pump Wisdom* because wisdom is defined as applied knowledge. If you concur with this very meaningful definition you will be ready for a rather serious challenge. That challenge is to practice wisdom by acting on the unique knowledge this text conveys.

Although both of us had written or co-authored other books and dozens of articles on centrifugal pump reliability improvement, some important material is too widely dispersed to be readily accessible. Moreover, some important material has never been published before. We again set out to assemble, rework, condense, and explain the most valuable points in a text aimed at ever wider distribution. With the addition of troubleshooting material, coverage of sensors, and centrifugal pump surveillance topics, it is now a text with, we hope, even more permanence and "staying power" than its predecessor edition. To thus satisfy the scope and intent of this book, we stayed with our original intention to keep theoretical explanations to a reasonable minimum and to limit the narrative to about 250 pages. Putting it another way, we wanted to squeeze into these 250 pages material and topics that will greatly enhance both centrifugal pump safety and reliability. All that is needed is the reader's solid determination to pay close attention and to follow up diligently.

Please again realize that in years past, many centrifugal pump manufacturers have primarily concentrated their design and improvement focus on the hydraulic end. Indeed, over time and in the decades since 1960, much advancement has been made in the metallurgical and power efficiency-related performance of the hydraulic assembly. Meanwhile, the mechanical assembly or drive-end of centrifugal process pumps was being treated with relative indifference. In essence, there was an imbalance between the attention given to pump hydraulics and pump mechanical issues.

Recognizing indifference as costly, this text will indeed rectify some of these imbalances. Our narrative and illustrations are intended to do justice to both the hydraulic assembly and the mechanical assembly of centrifugal process pumps. That said, the book briefly lays out how centrifugal pumps function and quickly moves to guidelines and details that must be considered by reliability-focused readers. A number of risky omissions or shortcuts by centrifugal pump designers, manufacturers, and users are also described.

The co-authors are indebted to AESSEAL. A worldwide manufacturer of sealing products, the company was unselfish in providing many images and special artwork without making us wait for the approval of layers of bureaucracy which, in a number of other instances, has delayed or even prevented the move to more reliable pump products. The company has demonstrated an exemplary respect for the environment and is often considered as an example of how a business should conduct itself for the mutual benefit of all parties.

Please take from us a strong measure of encouragement: Make good use of this book. Read it and apply it. Today, and hopefully years from now, remember to consult this material. Doing so will acquaint you and your successors with centrifugal pump failure avoidance and the more elusive aspects of preserving pump-related assets. And so, while you undoubtedly have more problems than you deserve, please keep in mind that you also deserve access to more solutions than you previously knew about or presently apply. Sound solutions are available, and they are here, right at your fingertips. Use them wisely; they will be cost-effective. The solutions you can discern from this text will have a positive effect on centrifugal pump safety performance and asset preservation. They have worked at Best-of-Class companies and cannot possibly disappoint you.

Robert X. Perez, P.E., and Heinz P. Bloch, P.E.
Fall 2020

1

Principles of Centrifugal Process Pumps

Pumps, of course, are simple machines that lift, transfer, or otherwise move fluid from one place to another. They are usually configured to use the rotational (kinetic) energy from an impeller to impart motion to a fluid. The impeller is located on a shaft; together, shaft and impeller(s) make up the rotor. This rotor is surrounded by a casing; located in this casing (or pump case) are one or more stationary passageways that direct the fluid to a discharge nozzle. Impeller and casing are the main components of the *hydraulic assembly*; the region or envelope containing bearings and seals is called the *mechanical assembly* or power end (Figure 1.1).

Many process pumps are designed and constructed to facilitate field repair. On these so-called "back pull-out" pumps, shop maintenance can be performed, while the casing and its associated suction and discharge piping (Figure 1.2) are left undisturbed. Although *operating* in the hydraulic end, the impeller remains with the power end during removal from the field.

The rotating impeller (Figure 1.3) is usually constructed with swept-back vanes, and the fluid is accelerated from the rotating impeller to the stationary passages into the surrounding casing.

In this manner, kinetic energy is added to the fluid stream (also called pumpage) as it enters the impeller's suction eye (A on Figure 1.3), travels through the impeller, and is then flung outward toward the impeller's periphery. After the fluid exits the impeller, it gradually decelerates to a much lower velocity in the stationary casing, called a volute casing, where the fluid stream's kinetic energy is converted into pressure energy (also called pressure head). The combination of the pump suction (inlet) pressure and the additional pressure head generated by the impeller creates a final pump discharge pressure that is higher than the suction pressure [3].

Pump Wisdom: Essential Centrifugal Pump Knowledge for Operators and Specialists,
Second Edition. Robert X. Perez and Heinz P. Bloch.
© 2022 The American Institute of Chemical Engineers, Inc. Published 2022 by John Wiley & Sons, Inc.

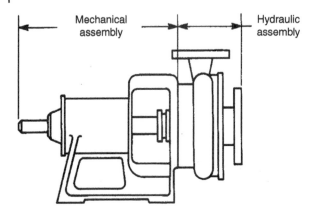

Hydraulic assembly

Impeller/propeller
Suction inlet
Volute
Seal rings

Mechanical assembly

Shaft seal

    Labyrinth
    Mechanical
    Packing

Shaft
Bearings
Housing/frame
Drive coupling/sheave

Figure 1.1 Principal components of an elementary process pump. *Source:* SKF USA, Inc. [1].

Pump Performance: Head and Flow

Pump performance is always described in terms of head H produced at a given flow capability Q, and hydraulic efficiency η attained at any particular intersection of H and Q. Head is customarily plotted on the vertical scale or vertical axis (the left of the two y-axes) of Figure 1.4; it is expressed in feet (or meters). Hydraulic efficiency is often plotted on another vertical scale, the right of the two vertical scales, i.e. the y-axis in this generalized plot.

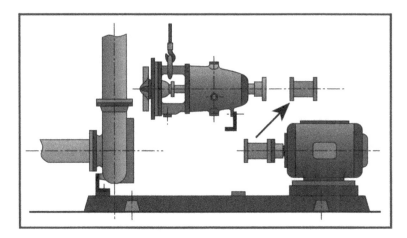

Figure 1.2 Typical process pump with suction flow entering horizontally and vertically oriented discharge pipe leaving the casing tangentially. *Source:* Emile Egger & Cie. [2].

Head is related to the difference between discharge pressure and suction pressure at the respective pump nozzles. Head is a simple concept, but this is where consideration of the impeller tip speed is important. The higher the shaft rpm and the larger the impeller diameter, the higher will be the impeller tip speed – actually its peripheral velocity.

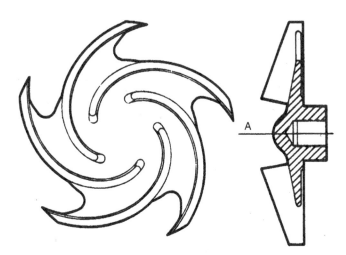

Figure 1.3 A semiopen impeller with five vanes. As shown, the impeller is configured for counterclockwise rotation about a centerline "A."

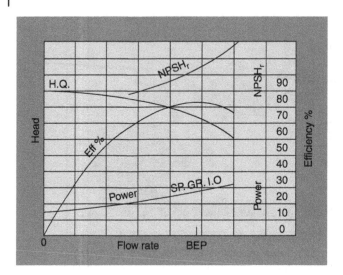

Figure 1.4 Typical "*H–Q*" performance curves are sloped as shown here. The best efficiency point (BEP) is marked with a small triangle; power and other parameters are often displayed on the same plot.

The concept of head can be visualized by thinking of a vertical pipe bolted to the outlet (the discharge nozzle) of a pump. In this imaginary pipe, a column of fluid would rise to a height "*H*". If the vertical pipe would be attached to the discharge nozzle of a pump with *higher* impeller tip speed, the fluid would rise to a greater height "*H*+". It is important to note that the height of a column of liquid, *H* or *H*+, is a function only of the impeller tip speed. The specific gravity of the liquid affects power demand but does not influence either *H* or *H*+. However, the resulting discharge pressure *does* depend on the liquid density (specific gravity or Sp.G.). For water (with an Sp.G. of 1.0), an *H* of 2.31 ft equals 1 psi (pound-per-square-inch), while for alcohol, which might have a Sp.G. of 0.5, a column height or head *H* of 4.62 ft equals 1 psi. So if a certain fluid had an Sp.G. of 1.28, a column height (head *H*) of 2.31/1.28 = 1.8 ft would equal a pressure of 1 psi.

For reasons of material strength and reasonably priced metallurgy, one usually limits the head per stage to about 700 ft. This is a fairly important rule-of-thumb limit to remember. When too many similar rule-of-thumb *limits* combine, one cannot expect pump reliability to be at its highest. As an example, say a particular impeller-to-shaft fit is to have 0.0002–0.0015 in. clearance on average size impeller hubs. With a clearance fit of 0.0015 in., one might anticipate a somewhat greater failure risk if this upper limit were found on an impeller operating with maximum allowable diameter.

On Figure 1.4, the point of zero flow (where the curve intersects the y-axis) is called the shut-off point. The point at which operating efficiency is at a peak is called the best efficiency point, or BEP. Head rise from BEP to shut-off is often chosen around 10–15% of differential head. This choice makes it easy to modulate pump flow by adjusting control valve open area based on monitoring pressure. Pumps "operate on their curves" and knowledge of what pressure relates to what flow allows technicians to program control loops.

The generalized depictions in Figure 1.4 also contain a curve labeled $NPSH_r$, which stands for Net Positive Suction Head required. This is the head of liquid that must exist at the edge of the inlet vanes of an impeller to allow liquid transport without causing undue vaporization. It is a function of impeller geometry and size and is determined by factory testing. $NPSH_r$ can range from a few feet to a three-digit number. At all times, the head of liquid *available* at the impeller inlet ($NPSH_a$) must exceed the required $NPSH_r$.

Operation at Zero Flow

The rate of flow through a pump is labeled Q (gpm) and is plotted on the horizontal axis (the x-scale). Note that for a given speed and for every value of head H we read off on the y-axis, there is a corresponding value of Q on the x-axis. This plotted relationship is expressed as "the pump is running on its curve." Pump H–Q curves are plotted to commence at zero flow and highest head. Process pumps need a continually rising curve inclination and a curve with a hump somewhere along its inclined line will not serve the reliability-focused user. Operation at zero flow is not allowed and, if over perhaps a minute's duration, could cause temperature rise and internal recirculation effects that might destroy most pumps.

But remember that this curve is valid only for this particular impeller pattern, geometry, size, and operation at the speed indicated by the manufacturer or entity that produced the curve. Curve steepness or inclination has to do with the number of vanes in that impeller; curve steepness is also affected by the angle each vane makes relative to the impeller hub. In general, curve shape is verified by physical testing at the manufacturer's facility. Once the entire pump is installed in the field, it can be re-tested periodically by the owner–purchaser for degradation and wear progression. Power draw may have been affected by seals and couplings that differ from the ones used on the manufacturer's test stand. Occasionally, high efficiencies are alluded to in a manufacturer's literature when bearing, seal, and coupling losses are not included in the vendor's test reports.

Impellers and Rotors

Regardless of flow classification centrifugal pumps range in size from tiny pumps to very big pumps. The tiny ones might be used in medical or laboratory applications; the extremely large pumps may move many thousands of liters or even gallons per second from flooded lowlands to the open sea.

All six of the impellers in Figure 1.5 are shown with a hub fastening the impeller to the shaft, and each of the first five impellers is shown as a hub-and-disc version with an impeller cover. The cover (or "shroud") identifies the first five as "closed" impellers; recall that Figure 1.3 had depicted a semiopen impeller. Semiopen impellers are designed and fabricated without the cover. Finally, open impellers come with free-standing vanes welded to or integrally cast into the hub. Since the latter incorporate neither disc nor cover, they are often used in viscous or fibrous paper stock applications.

To properly function, a semiopen impeller must operate in close proximity to a casing internal surface, which is why axial adjustment features are needed with these impellers. Axial location is a bit less critical with closed impellers. Except on axial flow pumps, fluid exits the impeller in the radial direction. Radial and mixed flow pumps are either single or double suction designs; both will be shown later. Once the impellers are fastened to a shaft, the resulting assembly is called a rotor.

In radial and mixed flow pumps, the number of impellers following each other, typically called "stages," can range from one to as many as will make such multistage pumps practical and economical to manufacture. Horizontal shaft pumps with up to 12 stages are not uncommon; using more than 12 stages on a horizontal shaft risks causing the rotor to resonate or vibrate at a so-called "critical speed." Vertical shaft pumps have been designed with 48 or more stages. In vertical

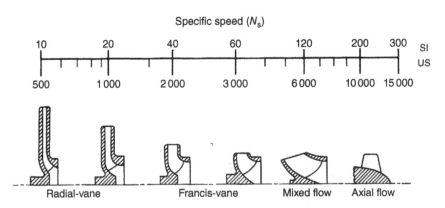

Figure 1.5 General flow classifications of process pump impellers.

pumps, shaft support bushings are relatively lightly loaded; they are spaced so as to minimize vibration risk.

The Meaning of Specific Speed

Pump impeller flow classifications and the general meaning of *specific speed* deserve further discussion. Moving from left to right in Figure 1.5, the various impeller geometries reflect selections that start with high differential pressure capabilities and end with progressively lower differential pressure capabilities. Differential pressure is simply discharge pressure minus suction pressure.

Specific speed calculations are a function of several impeller parameters; the mathematical expression includes exponents and is found later, in Figure 1.6. Staying with Figure 1.5 and again moving from left to right, we can reason that larger throughputs (flows) are more likely achieved by the configurations at right, whereas larger pressure ratios (discharge pressure divided by suction pressure) are usually achieved by the impeller geometries closer to the left of the illustration.

Impellers toward the right are more efficient than those near the left, and pump designers use the parameter *specific speed* (N_s) to bracket pump hydraulic efficiency attainment and other expected attributes of a particular impeller configurations and size. Please be sure not to confuse a very similar sounding parameter, *pump suction specific speed* (N_{ss} or N_{sss}), with the *specific speed* (N_s). For now, we are strictly addressing *specific speed* (N_s).

As an example, observe the customary use, whereby with N and Q – the typical given parameters that define centrifugal process pumps – one determines a pivot point. Next, with pivot point and head H, one can easily determine N_s. In Figure 1.5, Ns is somewhere between 500 and 15 000 on the US scale. Whenever we find ourselves in that range, we know such a pump exists, and we can even observe the general impeller shape. Keep in mind that thousands of impeller combinations and geometries exist. Impellers with covers are the most prominent in hydrocarbon processes, and an uneven number of impeller vanes is favored over even numbers of vanes for reasons of vibration suppression.

Pump specific speed, Figure 1.6, might be of primary interest to pump designers, but average users will also find it useful. On the lower right, the illustration gives the equation for N_s; it will be easy to see how N_s is related to the shaft speed N (rpm), flow Q (gpm or gallons/minute), and head H (expressed in feet). This mathematical expression also has two strange-looking exponents in it, but the N_s nomogram conveys more than meets the eye and can be quite helpful.

If now, we had the same or some other Q and would want to see what happens at some other speed, we would again draw a straight line to establish the pivot point. Drawing a line from whatever H is specified through the pivot

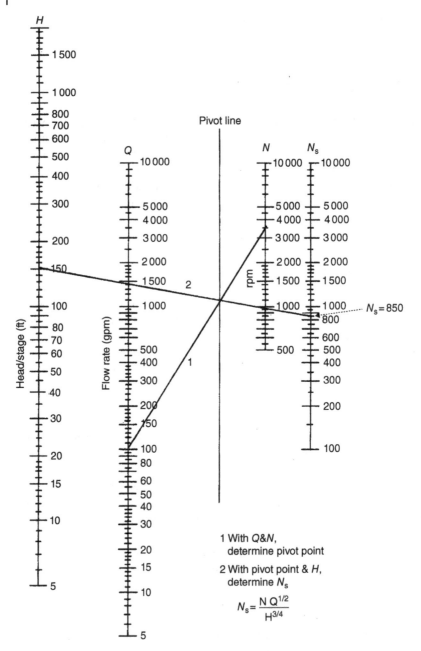

Figure 1.6 Pump specific speed nomogram allowing quick estimations. Shown in this illustration is a hypothetical pump application with a flow of 100 gpm operating at 3600 rpm (line 1). To develop 150 ft of head with a single stage, an impeller with a specific speed of 850 (line 2) would be required.

point and to N_s we would not like to select pumps with an N_s outside the rule-of-thumb range from 500 to 15 000. In another example, we might, after establishing the pivot point, wish to determine what happens if we select an impeller with the maximum head capability of 700 ft and draw a line through the pivot point. If the resulting N_s is too low, we would try a higher speed N and see what happens.

While there are always fringe applications in terms of size and flow rate, this book deals with centrifugal pumps in process plants. These pumps are related to the generic illustrations of Figures 1.1 and 1.2 and others in this chapter. All would somewhat typically – but by no means exclusively – range from 3 to perhaps 300 hp (2–225 kW).

Process Pump Types

The elementary process pumps illustrated in Figures 1.1 and 1.2 probably incorporate one of the radial vane impellers shown in Figure 1.5. If a certain differential pressure is to be achieved together with higher flows, such a pump is often designed with a double-flow impeller (Figure 1.7). One of the side benefits of double-flow impellers is very good axial thrust equalization (axial balance). A small thrust bearing will often suffice; it is shown here in the left bearing housing.

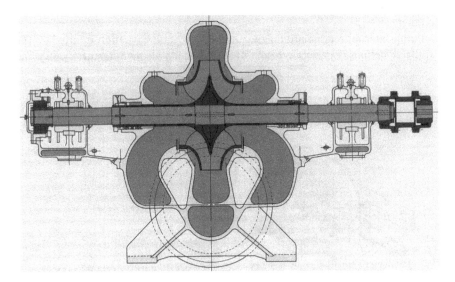

Figure 1.7 Double-flow impellers are used for higher flows and relatively equalized (balanced) axial thrust. *Source:* Mitsubishi Heavy Industries, Ltd. [4].

Note that the two radial bearings are plain, or sleeve-type. Certain sleeve bearings have relatively high speed capability.

If elevated pressures are needed, several impellers are lined up in series on the same pump rotor. Of course, this would then turn the pump into a multistage model Figure 1.8.

Process Pump Mechanical Response to Flow Changes

After the pumped fluid (also labeled pumpage, or flow) leaves at the impeller tip, it must be channeled into a stationary passageway that merges into the discharge nozzle. Many different types of passageway designs (single or multiple volutes, vaned diffusers, etc.) are available. Their respective geometry interacts with the flow and creates radial force action of different magnitude around the periphery of an impeller (Figure 1.9). These forces tend to deflect the pump shaft; they are greater at part-flow than at full flow.

Recirculation and Cavitation

Recirculation is a flow reversal near either the inlet or discharge of a centrifugal pump. This flow reversal produces cavitation-erosion damage that starts on the high-pressure side of an impeller vane and proceeds through the metal to the low-pressure side [5].

Pump-internal recirculation can cause surging and cavitation even when the available net positive suction head ($NPSH_a$) exceeds the manufacture's published

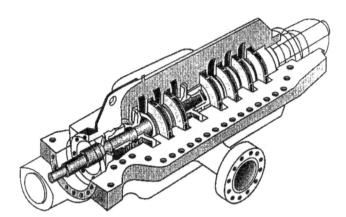

Figure 1.8 A multistage centrifugal process pump.

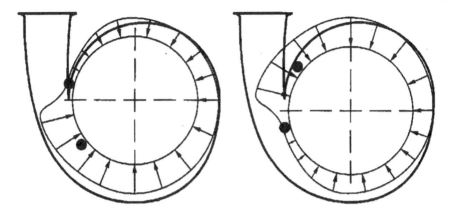

Figure 1.9 Direction and magnitude of fluid forces change at different flows. *Source:* World Pumps, February 2010, pp. 19.

NPSH required (NPSHr) by considerable margins. Also, extensive damage to the pressure side of impeller vanes has been observed in pumps operating at reduced flow rates. These are the obvious results of recirculation; however, more subtle symptoms and operational difficulties have been identified in pumps operating in the recirculation zone.

Symptoms of discharge recirculation are the following:

- Cavitation damage to the pressure side of the vane at the discharge
- Axial movement of the shaft, sometimes accompanied by damage to the thrust bearing
- Cracking or failure of the impeller shrouds at the discharge
- Shaft failure on the outboard end of double-suction and multistage pumps
- Cavitation damage to the casing tongue (see Chapter 11) or diffuser vanes

Symptoms of suction recirculation are the following:

- Cavitation damage to the pressure side of the vanes at the inlet
- Cavitation damage to the stationary vanes in the suction
- Random crackling noise in the suction; this contrasts with the steady crackling noise caused by inadequate net positive suction head
- Surging of the suction flow

A quick-reference illustration was provided by Warren Fraser Figure 1.10. We should note that recirculation and the attendant failure risks are low in pumps delivering 2500 gpm or less at heads up to 150 ft. In those pumps, energy levels may not be high enough even if the pump operates in the recirculation zone. As a general rule for such pumps, minimum flow can be set at 50% of recirculation

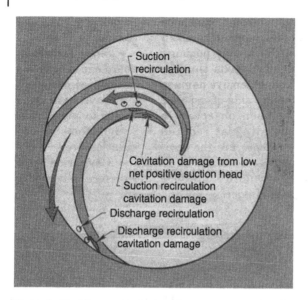

Figure 1.10 Where and why impeller vanes get damaged. *Source:* Fraser [5].

flow for continuous operation and 25% of recirculation flow for intermittent operation [6].

The Importance of Suction Specific Speed

Note that pump suction specific speed (N_{ss} or N_{sss}) differs from the pump specific speed N_s discussed earlier. For installations delivering over 2500 gpm and with suction specific speeds over 9000, greater care is needed. Suction specific speed (N_{sss} or N_{ss}) is calculated by the straightforward mathematical expression:

$$N_{ss} \frac{\left(r/\min \right)\left[\left(gal/\min \right)/eye \right]^{1/2}}{\left(NPSH_r \right)^{3/4}}$$

wherein both the flow rate and $NPSH_r$ pertain to conditions published by the manufacturer. In each case, these conditions (flow in gpm) and $NPSH_r$ are observed on the maximum available impeller diameter for that particular pump.

The higher the design suction specific speed, the closer the point of *suction* recirculation to what is commonly described as rated capacity. Similarly, the closer the *discharge* recirculation capacity is to rated capacity, the higher the efficiency. Pump system designers are tempted to aim for highest possible efficiency and suction specific speed. However, such designs might result in systems with either limited pump operating range or, if operated inside the recirculation range, disappointing reliability and frequent failures.

Although more precise calculations are available, trend curves of probable NPSHr for minimum recirculation and zero cavitation-erosion in water, Figure 1.11, are sufficiently accurate to warrant our attention [7]. The NPSHr needed for *zero* damage to impellers and other pump components may be many times that published in the manufacturer's literature. The manufacturers' NPSHr plot (lowermost curve in Figure 1.11) is based on observing a 3% drop in discharge head or pressure; at $Q = 100\%$, we note NPSHr = 100% of the manufacturer's claim. Unfortunately, whenever this 3% fluctuation occurs, a measure of damage may already be in progress. Assume the true NPSH<u>r</u> is as shown in Figure 1.11 and aim to provide an $NPSH_a$ in excess of this true NPSHr.

In Ref. [7], Irving Taylor compiled his general observations and alerted us to this fact. He cautioned against considering his curves totally accurate and mentioned the demarcation line between low and high suction specific speeds somewhere between 8000 and 12 000. Many data points taken after 1980 point to 8500 or 9000 as numbers of concern. If pumps with N_{ss} numbers higher than 9000 are being operated at flows much higher or lower than BEP, their life expectancy or repair-free operating time will be reduced.

In the decades after Taylor's presentation, controlled testing has been done in many industrialized countries. The various findings have been reduced to relatively accurate calculations that were later published by HI, the Hydraulic

Trend of probable NPSH$_r$ for zero cavitation-erosion

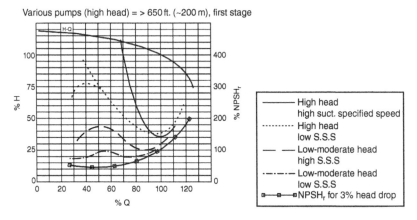

Figure 1.11 Pump manufacturers usually plot only the NPSHr trend associated with the lowermost curve. At that time a head drop or pressure fluctuation of 3% exists and cavitation damage is often experienced. *Source:* Taylor [7].

Institute [8]. Relevant summaries can also be found in Ref. [9]. Calculations based on Refs. [8, 9] determine minimum allowable flow as a percentage of BEP.

Note, again, that recirculation differs from cavitation, a term which essentially describes vapor bubbles that collapse. Cavitation damage is often caused by low net positive suction head available ($NPSH_a$). Such cavitation-related damage starts on the low-pressure side and proceeds to the high-pressure side. An impeller requires a certain net positive suction head; this $NPSH_r$ is simply the pressure needed at the impeller inlet (or eye) for relatively vapor-free flow.

What We Have Learned

Understanding the concepts of N_s and N_{ss} will assist in specifying better pumps. In addition to fluid properties pump life is influenced by throughput. Just as an automobile transmission is designed to work best at particular speeds or optimum gear ratios, pumps have desirable flow ranges. Deviating from optimum flow will influence failure risk and life expectancy.

References

1 SKF USA, Inc.; "Bearings in Centrifugal Pumps", Kulpsville, PA, Publication 100-955, Version 4/2008; excerpted or adapted by permission of the copyright holder.
2 Emile Egger & Cie.; "Operating Manual", Salt Lake City, Utah, and Cressier, NE, Switzerland, 2008.
3 ITT/Goulds Pump Corporation; "Installation and Maintenance Manual for Model 3196 ANSI Pump", Seneca Falls, NY, 1990.
4 Mitsubishi Heavy Industries, Ltd.; "Publication HD30-04060", Tokyo, Japan, and New York, NY.
5 Fraser, Warren H.; "Flow recirculation in centrifugal pumps," Proceedings of the Texas A&M University Turbomachinery Symposium, Houston, TX, 1981, pp. 95–100.
6 Ingram, James H.; "Pump reliability–where do you start", presented at ASME Petroleum Mechanical Engineering Workshop and Conference, Dallas, TX, September 13–15, 1981.
7 Taylor, Irving; "The Most Persistent Pump-Application Problems for Petroleum and Power Engineers", ASME Publication 77-Pet-5 (Energy Technology Conference and Exhibit, Houston, Texas, September 18–22, 1977).
8 ANSI/HI9.6.3-1997; "*Allowable Operating Region*", Hydraulic Institute, Parsippany, New Jersey, 2008.
9 Bloch, Heinz P., and Alan Budris; "*Pump User's Handbook*", 3rd Edition, Fairmont Press, Lilburn, GA 30047, 2010 (ISBN 0-88173-517-5).

2

Pump Selection and Industry Standards

Anybody can buy a cheap pump, but you want to buy a better pump. The term "better pumps" describes fluid movers that are well designed beyond just hydraulic efficiency and modern metallurgy. Better pumps are ones that avoid risk areas in the mechanical portion commonly called the drive-end. That is the part of process pumps that has been neglected most often and where cost cutting should cause the greatest concern.

Deviations from best available technology increase the failure risk. As three or four or more deviations combine, a failure is very likely to occur. An analogy could be drawn from an incident involving two automobiles, with one driving behind the other. When the trailing vehicle travelled at (i) an excessive speed, with (ii) worn tires, on (iii) a wet road, and (iv) followed the leading car too closely, a rear-end collision resulted. Had there just been any three of the four violations, the event might be recalled as one of the many "near miss" incidents. Had there been any two of the four, it would serve no purpose to tell the story in the first place.

Too much cost-cutting by pump manufacturers and purchasers will negatively affect the drive-ends of process pumps. Flawed drive-end components are therefore among the main contributors to elusive repeat failures that often plague pumps – essentially very simple machines. Drive-end flaws deserve to be addressed with urgency, and this short chapter will introduce the reader to more details that follow later in Chapters 5, 8 and 10.

Why Insist on Better Pumps

Well-informed reliability professionals will be reluctant to accept pumps that incorporate the drive-end shown in Figure 2.1. The short overview of reasons is that reliability-focused professionals take seriously their obligation to consider the

Pump Wisdom: Essential Centrifugal Pump Knowledge for Operators and Specialists,
Second Edition. Robert X. Perez and Heinz P. Bloch.
© 2022 The American Institute of Chemical Engineers, Inc. Published 2022 by John Wiley & Sons, Inc.

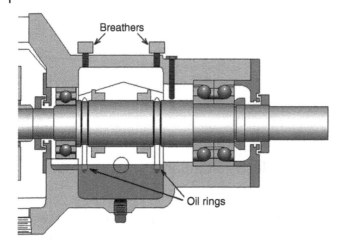

Figure 2.1 A typical bearing housing with several potentially costly vulnerabilities.

actual, *lifetime-related* and not just *short-term*, cost of ownership. They have learned long ago that price is what one pays, and value is what one gets.

Anyway, while at first glance the reader might see nothing wrong, Figure 2.1 contains clues as to why many pumps fail relatively frequently and sometimes quite randomly. It shows areas of vulnerability that must be recognized and eliminated. The best time to eliminate flaws is in the specification process. A number of important vulnerabilities, deviations from best available technology or just plain risk areas exist in that illustration:

- Oil rings are used to lift oil from the sump into the bearings;
- The back-to-back oriented thrust bearings are not located in a cartridge;
- Bearing housing protector seals are missing from this picture;
- Although the bottom of the housing bore (at the radial bearing) shows the desired passage, the same type of oil return or pressure equalization passage is *not* shown near the 6 o'clock position of the thrust bearing;
- There is uncertainty as to the type or style of constant level lubricator that will be supplied. Unless specified, the pump manufacturer will almost certainly provide the least expensive constant level lubricator configuration. Putting it another way: The best lubricators are rarely found on newly sold pumps.

Each of these issues merits further explanation and will be discussed in Chapters 5 through 9. Recall also that our considerations are confined to lubrication issues on process pumps with liquid oil-lubricated rolling element bearings. The great majority of process pumps used worldwide belong to this lubrication and bearing category. Small pumps with grease-lubricated bearings and large pumps with sleeve bearings and circulating pressure-lube systems are not discussed in this text.

ANSI and ISO vs. API Pumps

ANSI stands for American National Standards Institute; ISO is the International Standards Institute, and API is the American Petroleum Institute. In general, ANSI and ISO pumps comply with dimensional standards; the measurement conventions are inches and millimeters, respectively. ANSI pumps will have the same principal dimensions regardless of manufacturer, as will ISO-compliant pumps in their respective size groups. Principal dimensions, for the sake of this overview include, but are certainly not limited to, the distance from base mounting surfaces to the shaft centerline, or to the pump suction and discharge flange faces.

The widely used API-610 pump standard is aiming for high strength and reliability. This standard is often used for pumping services that qualify for one or more of the labels hazardous, flammable, toxic, or explosion-prone. The API-610 standard has been called a quality standard, although that should not be viewed as a negative comment on the fitness of ANSI and ISO pumps for safe and long-term satisfactory service. API pumps are centerline-mounted (Figure 2.2); ANSI pumps are usually foot-mounted (as shown earlier in Figures 1.1 and 1.2). The thermal rise of the shaft centerline of a horizontal ANSI pump can be as much as

Figure 2.2 The suction and discharge nozzles on this centerline-mounted API-style pump are upward-oriented. Oil mist lubrication is applied throughout; the coupling guard was removed for maintenance. *Source:* Lubrication Systems Company [1].

three times greater than that of the centerline-mounted API pump. Thermal rise is taken into account during pump alignment (Chapter 14).

But the API-610 standard should not be viewed as infallible and the wording in the inside cover of the standard makes that often-overlooked point. Reliability-focused users have seen justified to deviate from it when experience and technical justification called for such deviations. This text will deal with some of the issues where API-610 needs user attention and suitable amendment.

Experience-based selection criteria summarize the proven practices of the Monsanto Chemical Company's Texas City plant in the 1970s [2]. Among these was the recommendation of using in-between-bearing pump rotors whenever the product of power input and rotational speed (kW times rpm) would exceed 675 000.

For general guidance [2] asked that API-610 compliant pumps be given strong consideration whenever one or more of the following six conditions are either reached or exceeded:

- Head exceeds 350 ft (~106 m);
- Temperature exceeds 300 °F (~150 °C) on pipe up to and including 6 in. nominal diameter; alternatively, if the temperature exceeds 350 °F (~177 °C) on pipe starting with 8 in. nominal diameter;
- Pumps with drivers rated in excess of 100 hp (starting at 75 kW and higher);
- Suction pressures over 75 psig (516 kPa);
- Flow in excess of the flow at best efficiency point (BEP) for the pump at issue;
- Speeds in excess of 3600 rpm.

Exceptions to the six conditions can be made judiciously. To qualify for such an exception, the pumped fluid should be nonflammable, nontoxic, and nonexplosive. In general, exceptions might be granted if the vendor can demonstrate years of successful operation for the proposed pump in a comparable or perhaps even more critical service.

Best-in-class (BiC) pump users are ones that are able to get long failure-free runs from their pumps. BiCs have on their bidders' lists only vendors (or custom builders) with proven experience records. Such vendors and manufacturers would be well established and would have a record of sound quality and on-time deliveries.

Exceptions taken by a bidder to the owner-operator's specification would be carefully examined for their potential reliability impact. This examination process serves as a check on the pump manufacturers' understanding of the buyer's long-term reliability requirements. Orders would be placed with competent bidders only.

Because these vendors use a satisfied workforce of experienced specialists, do effective training and mentoring, and have not disbanded their quality control and inspection departments, their products will command reasonable pricing. Reasonable pricing should not be confused with lowest pricing, although reasonable pricing may indeed be lowest in terms of life cycle costing.

Vertical pumps are available in many hundreds of styles and configurations and Figure 2.3 shows a two-stage pump custom-built for pipeline service. This pump is unique because the entire pumping element can be removed as one piece for maintenance. The motor mount (with the motor attached) would be removed first, 24 screws are removed next, and the whole pump lifted out. It is a good example of a design that is user-oriented in terms of maintenance and probable overall reliability [3].

Both API and non-API standards are used in custom-built pumps, depending on user preference, type of service, and prevailing experience. Competent designs are available not only from original equipment manufacturers (OEMs) but also from certain key custom design innovators and manufacturing specialists. We count them among the quality providers.

In all instances, the pump owner-operator would compile a specification document that incorporates most, if not all, of the items discussed in this text. The pump owner-operator or its designated project team would mail the document to at least two, but more probably three or four of these quality bidders or quality providers. Their replies or cost quotes would be carefully reviewed. These replies would describe the vendor's offer pictorially, and Figures 2.4 and 2.5 are typical of

Figure 2.3 Custom-built vertical pipeline pump with drive motor removed. *Source:* Alfred Conhagen Inc. [3].

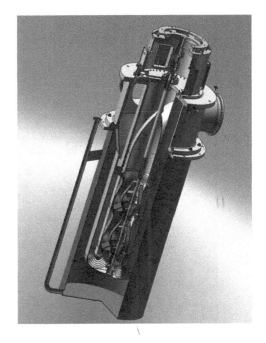

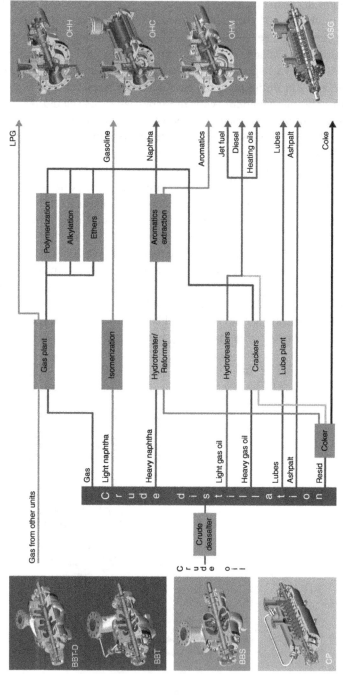

Figure 2.4 HPI process scheme and API pumps offered for the various services. *Source:* Sulzer Pumps, Ltd. [4].

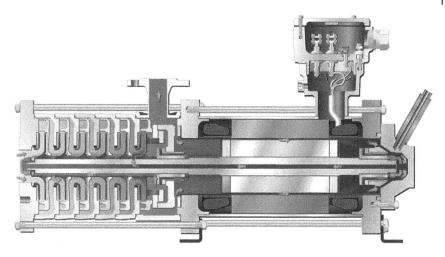

Figure 2.5 Canned Motor Pumps can be superior alternatives to conventional centrifugal process pumps. *Source:* Hermetic Pumpen, Gundelfingen, Germany.

add-ons and/or alternatives that the vendor can submit together with suitable documents in support of its claim of proven experience.

What We Have Learned

- Getting good pumps requires an up-front effort of defining what the buyer really wants. The user's present or future maintenance philosophy will determine what belongs into a specification. The bid invitation must encourage vendor-manufacturers to alert owner-purchasers to superior choices, if available.
- Without a good specification, the buyer is very likely to get a "bare minimum" product. Bare minimum products will require considerable maintenance and repair effort in future times. For instance, oil mist lubrication is not usually included in "bare minimum" offers [5].
- The specification document must be submitted to competent bidders only. Some bidders may ask you to grant a waiver to a particular specification clause; insist they prove that they have understood the reason and purpose of the clause they are unable or unwilling to meet. Never waste your time on bidders that take blanket exception to your entire specification.
- If the owner–purchaser of a process plant grants a waiver to a certain specification clause, he should understand the extent to which noncompliance will lead to increased maintenance requirements, downtime, or even catastrophic failure risk.
- In the end you get not what you *ex*pect, but you get what you *in*spect. Inspection is one of the costs of getting reliable process pumps.

References

1 Lubrication Systems Company; "Photo contributed by and used with the permission of Don Ehlert", Houston, Texas, 2008.

2 Ingram, J.H.; "Pump reliability – Where Do You Start", presented at ASME Petroleum Mechanical Engineering Workshop and Conference, Dallas, TX, September 13–15, 1981.

3 Alfred Conhagen Inc.; Houston, Texas, 2010.

4 Sulzer Pumps, Ltd.; Winterthur, Switzerland. By permission, 2010.

5 Bloch, Heinz P.; *"Optimized Equipment Lubrication: Conventional Lubrication, Oil Mist Technology, and full Standby Protection"*, DeGruyter Publishing, Berlin/ Germany, 2021 (ISBN 978-3-11-074934-2)

3

Foundations and Baseplates

Pumps can be found mounted in many different ways; there are times and places to do it at least cost and times and places to do it with uncompromisingly high quality.

Plants that use stilt mounting (Figure 3.1) often fall short of achieving best-possible equipment reliability. Best practices plants secure their pumps more solidly on more traditional foundations. Stilt-mounted pump sets lack overall stiffness but have been used for small ANSI pumps where the sideways-move capability of the entire installation was thought to equalize piping-induced stresses. Among its few advantages is low initial cost.

However, there are serious shortcomings since stilt-mounting will not allow pump vibration to be transmitted through the baseplate to the foundation and down through the subsoil. Proper foundation-mounting permits transmission of vibration which can result in a significant increase in mean time between failures (MTBFs), longer life of mechanical seals and bearings, and favorably low total life-cycle cost [1].

Securing Pumps in Place – With One Exception

Again, proper field installation of pumps has a measurable positive impact on pump life. Even a superb design will give poor results if poorly installed. A moderately good pump design, properly installed, will give good results [2]. Proper installation refers to a good foundation design, no pipe strain (see Chapter 4), and good shaft alignment (Chapter 14) to name just a few. No pump manufacturer designs its pumps strong enough to act as a solid anchoring point for incorrectly supported piping, or piping that causes casings and pump nozzles to yield and deflect. Also, pumps have to be properly secured to their respective baseplates,

Pump Wisdom: Essential Centrifugal Pump Knowledge for Operators and Specialists, Second Edition. Robert X. Perez and Heinz P. Bloch.
© 2022 The American Institute of Chemical Engineers, Inc. Published 2022 by John Wiley & Sons, Inc.

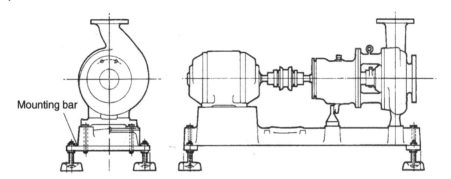

Mounting bar

Figure 3.1 ANSI pump set on a stilt-mounted baseplate. *Source:* ITT/Goulds, Seneca Falls, NY.

and these baseplates have to be well-bonded to the underlying foundation. Epoxy grout is used to do this bonding in modern installations.

There is one exception, however. Vertical in-line pumps (Figure 3.2) are not to be bolted to the foundation. They are intended to respond to thermal and other growths of the connected piping and must be allowed to float or slide a fraction of an inch in the x- and y-directions. The foundation mass under vertical in-line pumps can be much less than that under the more typical horizontal pump.

Making the foundation mass three to five times the mass of the pump and its driver has been the rule of thumb for horizontal pumps. For vertical in-line

 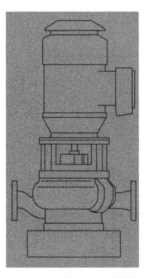

Figure 3.2 Vertical in-line pumps are not to be bolted to the foundation. They should be allowed to move with the connected pipes.

pumps, it is acceptable to make the concrete foundation about one-and-a-half to twice the mass of the pump-and-driver combination [3].

Why Not to Install Pump Sets in the As-Shipped Condition

There are obviously some flaws in the grout surrounding the baseplate in Figure 3.3. (On the other hand, the equipment owner invested in a very modern small oil mist lubrication unit). Note the hollow space under the electric motor. Lack of support under motors often invites resonant vibration. Rigorous written installation procedures are needed and must be adhered to if long equipment life is to be achieved.

Before delving into other installation matters, note the alignment jacking provisions in Figure 3.4, where the purchaser specified an arrangement that allows insertion (and later removal) of alignment-jacking tabs in the *x*- and *y*-directions next to each of the four motor feet. (A fixed jacking tab arrangement can be seen later – Figure 14.3).

Portable jacking tabs, Figure 3.4, (inserted in a welded-on bracket), allow driver alignment moves to be made. Thereafter, the jacking bolts are backed-off, and the entire tab is removed. When jack screws are left tightened against the motor feet, motor heat, and thermal growth might force the feet into these

Figure 3.3   A typical, but obviously flawed, "conventional" pump foundation. *Source:* Lubrication Systems Company, Houston, Texas.

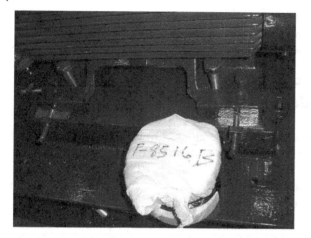

Figure 3.4 Removable alignment jacking tabs shown inserted in three of four locations next to the two motor feet shown here. *Source:* Stay-Tru®, Houston, Texas.

bolts even more, sometimes causing the entire motor casing to distort [1]. Note, therefore, that backing-off jacking bolts should be one of many installation checklist items.

To ensure level-mounting throughout, the baseplate is placed on a foundation into which hold-down bolts or anchor bolts (Figure 3.5) were encased when the reinforced concrete foundation was being poured [2]. For proper stretch and long life, these anchor bolts (Figure 3.5) must have a diameter-to-length ratio somewhere between 1 : 10 and 1 : 12. The anchor bolts are provided with steel sleeves and soft filler. The sleeves prevent entry of grout and accommodate the differing amounts of thermal growth of a concrete foundation relative to that of a steel baseplate.

Conventional vs. Prefilled Baseplate Installations

In general, horizontal process pumps and drivers are shipped and received as a "set" or package, i.e. already premounted on a baseplate. Seeing a

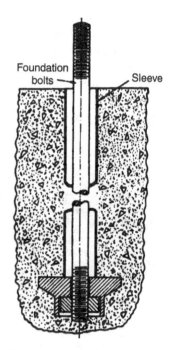

Figure 3.5 Foundation anchor bolts and sleeves encased in pump foundation. *Source:* Based on Barringer and Monroe [2].

conveniently mounted-for-shipping pump set very often leads to the erroneous assumption that the entire package can simply be hoisted up and placed on a suitable foundation. However, that's certainly not best practice and best-in-class (BiC) plants will not allow it.

Shipping method has little to do with how pumps should best be installed in the field, and pump installation issues merit considerable attention. Again, before installing a *conventional* baseplate, the pump and its driver must be removed from the baseplate and set aside. Leveling screws (Figure 3.6) are then used in conjunction with laser-optic tools or a machinist's precision level. With the help of these tools, the baseplate-mounting pads are brought into flat and parallel condition side-to-side, end-to-end, and also diagonally, all within an accuracy of 0.001 in./ft (~0.08 mm/m) or better. The nuts engaging the anchor bolts are being secured next, and the hollow spaces within the baseplate as well as the space between baseplate and foundation filled with epoxy grout.

The traditional approach to joining the baseplate to the foundation has been to build a liquid-tight wooden form around the perimeter of the foundation and fill the void between the baseplate and the foundation with either a cement-based or epoxy grout. Both grouting approaches are considered conventional and should not be confused with the preferred epoxy prefilled method which is highlighted below.

Grouting a baseplate or skid to a foundation requires careful attention to many details. A successful grout job will provide a mounting surface for the equipment that is flat, level, very rigid, and completely bonded to the foundation system. Many times these attributes are not obtained during the first attempt at grouting, and expensive field correction techniques have to be employed. Predominant

Figure 3.6 Steel baseplate with anchor bolt shown on left and leveling screw on right. A chock (thick steel washer) is shown between leveling screw and foundation [2]. *Source:* Modern Pumping Today.

installation problems involve voids and distortion of the mounting surfaces. In fact, the most frequently overlooked foundation and pipe support problems are related to foundation settling.

How so? Just as a residential dwelling or sidewalk will probably shift, settle, and crack over time, pump foundations and supports should be expected to do the same. It would be prudent to plan for preventive or corrective action over time or during plant shutdowns. Fortunately, there is now an even better option; it involves the use of standard baseplates prefilled with epoxy.

Epoxy Prefilled Baseplates

As of about 2000, Best Practices Companies (BPCs) have increasingly used "monolithic" (all-in-one, epoxy prefilled) steel baseplates in sizes approaching 1.5 m × 2.5 m (about 5 ft × 8 ft). Larger sizes become cumbersome due to heavy weight.

In the size range up to about 1.5 m × 2.5 m, conventional grouting procedures, although briefly mentioned in this text, are being phased out in favor of baseplates prefilled with an epoxy resin or grout [3]. These standard material prefilled steel baseplates then represent a solid block (the "monolith") that will never twist and never get out-of-alignment.

The process includes five successive stages, all done under controlled conditions before shipment to the site:

1) Baseplate fabrication. (No pour holes are needed for prefilled baseplates)
2) Stress relieving
3) Pregrouting (primer application) in preparation for prefilling. (If there are large pour holes, the inverted baseplate must be placed on a sheet of plywood, Figure 3.7)
4) Fill with epoxy grout and allow it to bond and cure
5) Invert and machine the mounting pads to be flat; then verify flatness before shipment (Figure 3.8). Protect and ship (Figure 3.9) – possibly even with pump, coupling, and driver-mounted and final-aligned.

The advantage of prefilling is notable. Jobs with pumps in the 750 kW (1000 hp) category and total assemblies weighing over 12 000 kg (26,400 lbs) have been done without difficulty on many occasions. A conventionally grouted baseplate requires at least two pours, plus locating and repair-filling of voids after the grout has cured. Prefilled or pregrouted baseplates travel better and arrive at the site flat and aligned, just as they left the factory. Their structural integrity is better because they do not require grout holes. Their installed cost is less and their long-term reliability is greatly improved.

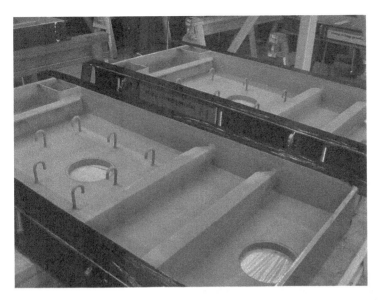

Figure 3.7 Underside of a baseplate after a prime coat has been applied. It is ready to be filled with epoxy. The large pour holes identify it as an old-style "conventional" baseplate being converted to prefilled style. *Source:* Stay-Tru®, Houston, TX.

Figure 3.8 Flatness and level measurements determine if the now-machined-prefilled baseplate has been properly machined. It is then ready to be installed on a foundation at site. *Source:* Stay-Tru®, Houston, Texas.

Figure 3.9 Epoxy prefilled baseplate fully manufactured by a specialty company, shown ready for shipment. *Source:* Stay-Tru® Company, Houston, TX.

How to Proceed If There Is No Access to Specialist Firms

If a specialist firm is not available or if upgrading is done at a field location, ascertain that the baseplate's underside is primed with high-quality epoxy paint. In general, baseplates are specified with an epoxy primer on the underside. This primed underside should be solvent-washed, lightly sanded to remove the glossy finish, and solvent-washed again. For inorganic zinc and other primer systems, the bond strength to the metal should be determined; expert instructions will be quite helpful. There are several methods for determining bond strength, but, as a general rule, if the primer can be scraped off with a putty knife, the primer should be removed. Sand blasting to an SP-6 finish is the preferred method for primer removal. After sand blasting, the surface should first be solvent-washed and then pregrouted (i.e. epoxy-filled) within eight hours.

By its very nature, pregrouting a baseplate will greatly reduce problems of entrained air-creating voids. However, because grout materials are highly viscous, proper placement of the grout is still important to prevent air pockets from developing [4]. The baseplate must also be well supported (Figure 3.7) to prevent severe distortion of the mounting surfaces due to the weight of the grout.

Once the pregrouted baseplate has been fully cured, it is turned right-side-up and a complete inspection of the mounting surfaces is performed (Figure 3.8). If surface grinding of the mounting surfaces is necessary, then a postmachining inspection must also be performed. Careful inspection for flatness, coplanarity, and relative levelness (colinear) surfaces should be well documented for the facility's construction or equipment files. The methods and tolerances for inspection should conform to the following:

Flatness: A precision ground parallel bar is placed on each mounting surface. The gap between the precision ground bar and the mounting surface is measured with a feeler gauge. The critical areas for flatness are within a 2–3″ radius of the equipment hold-down bolts. Inside of this area, the measured gap must be less than 0.001″. Outside the critical area, the measured gap must be less than 0.002″. If the baseplate flatness falls outside of these tolerances, the baseplate needs to be surface ground.

Coplanarity: A precision ground parallel bar is used to span across the pump and motor mounting pads in five different positions, three lateral and two diagonal. At each location, the gap between the precision ground bar and the mounting surfaces is measured with a feeler gauge. If the gap at any location along the ground bar is found to be more than 0.002″, then the mounting pads will be deemed noncoplanar and the baseplate will have to be surface ground (Figure 3.8). Outsourcing baseplate design, fabrication, and prefilling with epoxy grout has often been found economically attractive. Figure 3.9 shows it ready for shipment.

Relative level (Colinearity): It is important to understand the difference between relative level and absolute level. Absolute level is the relationship of the machined surfaces to the earth. The procedure for absolute leveling is done in the field and is not a part of this inspection. Relative level is an evaluation of the ability to achieve absolute level before the baseplate gets to the field.

Conventional grouting methods for nonfilled baseplates, by their very nature, are labor and time intensive [5]. Utilizing a pregrouted baseplate with conventional grouting methods helps to minimize some of the cost, but the last pour still requires a full grout crew, skilled carpentry work, and good logistics. To further minimize the costs associated with baseplate installations, a new field grouting method has been developed for pregrouted baseplates. This new method [6] utilizes a low-viscosity, high-strength epoxy grout system that greatly reduces foundation preparation, grout form construction, crew size, and the amount of epoxy grout used for the final pour.

What We Have Learned: Checklist of Foundation and Baseplate Topics

- Use ultrastiff, epoxy-filled formed steel baseplates ("StayTru*" method or an approved equivalent) on new projects and on optimizing existing facilities
  - Proceed by first inverting and preparing the baseplate; use recommended grit blasting and primer paint techniques.
  - Fill with suitable epoxy grout to become a monolithic block.

- Allow to cure; after curing, turn over and machine all mounting pads flat and coplanar within 0.0005 in./ft (0.04 mm/m).
- Next, install complete baseplate on pump foundation. Anchor and level it within the same accuracy.
- At final installation, place epoxy grout between the top of the foundation and the space beneath the monolithic epoxy prefilled baseplate.
- On welded baseplates, make sure that standard thickness steel is used and that the welds are continuous and free of cracks.
- On pump sets with larger than 75 kW (100 hp) drivers, ascertain that baseplates are furnished with eight positioning screws per casing, i.e. two screws ("jacking bolts") per mounting pad:
  - These positioning screws could be located in removable tabs (i.e. tabs slipped into a welded guide bracket) or fixed tabs (i.e. tabs welded onto the baseplate).
  - Pad heights must be such that at least 1/8 in. (3 mm) stainless steel shims can be placed under driver feet.
- Conventional baseplates must be installed and grouted on foundation with pump and driver removed. Only then should pump and driver be reinstalled and leveled.
- Epoxy-prefilled baseplates can be installed and grouted on a foundation with pump and driver already aligned and bolted down on the baseplate.

References

1 Myers, R.; "Repair Grouting to Combat Pump Vibration", *Chemical Engineering Book Series*, Volume 1, McGraw-Hill, New York, NY, 1998.
2 Barringer, Paul, and Monroe, Todd; "How to Justify Machinery Improvements Using Reliability Engineering Principles", Proceedings of the Sixteenth International Pump Users Symposium, Turbomachinery Laboratory, Texas A&M University, College Station, Texas, 1999.
3 Bloch, Heinz P., and Budris, A. R.; *"Pump User's Handbook: Life Extension"*, 3rd Edition, Fairmont Press, Lilburn, GA, 2010 (ISBN 0-88173-627-9).
4 Bloch, Heinz P., and Geitner, F. K.; *Major Process Equipment Maintenance and Repair"*, 2nd Edition, Gulf Publishing Company, Houston, TX, 1990 (ISBN 0-88415-663-X).
5 Harrison, Donald M.; *"The Grouting Handbook"*, Gulf Publishing Company, Houston, TX, 2000 (ISBN 0-88415-887-X).
6 Monroe, Todd R., and Palmer, Kermit L.; *"Methods for the Design and Installation of Epoxy Pre-filled Base Plates"*, Marketing Bulletin, Stay-Tru® Services, Inc., Houston, Texas, 1997.

4

Piping, Stationary Seals, and Gasketing

Pipe Installation and Support

Unless piping configurations and associated layout are done by thoughtful design, things often go wrong. A distinction must be made between the implementation tasks assigned to pipe fitters and the reliability-focused engineering tasks assigned to piping designers.

Accurate computer models and suitable software are used by reliability-focused plants. These plants then obtain the lowest stress, most cost-effective, and least-risk pipe installation by design, and not by any of the less-dependable means.

Based on computer recommendations, fixed and sliding pipe supports are located so as not to let thermal movements cause undue loads on pump nozzles. At some locations, the pipe must be suspended by tension springs; at other locations, the pipe must rest on suitably restrained compression springs. The weight of insulation and sometimes even ice or snow loads must be taken into account by the designer.

Sliding Supports and Installation Sequence Deserve Special Attention

There is a widely overlooked item on sliding supports: Steel-Teflon®-Steel is not usually a satisfactory sliding support. The steel plates ("shoes") will oxidize and, due to surface roughness, will dig into the Teflon. Best practice is to use Steel-Teflon®-Teflon®-Steel, in which case Teflon® will slide on Teflon® with considerable ease, providing long-term satisfactory sliding action.

Pump Wisdom: Essential Centrifugal Pump Knowledge for Operators and Specialists, Second Edition. Robert X. Perez and Heinz P. Bloch.
© 2022 The American Institute of Chemical Engineers, Inc. Published 2022 by John Wiley & Sons, Inc.

Best practice requires installation of piping from vessels or other structures upstream of the pump and toward the pump suction nozzle, to a point about 10–15 ft (3–5 m) from the suction nozzle. Then, one places a pipe flange at the suction nozzle and works toward the pipe run that had been terminated 10–15 ft upstream. The same work sequence is next utilized on the downstream piping.

On the downstream pipe installation, one would install pipe from receivers, destination vessels, or other structures downstream of the pump and install pipe toward the pump discharge nozzle. This downstream pipe installation sequence should initially terminate at a point about 10–15 ft (3–5 m) from the pump discharge nozzle. Then, one places a pipe flange at the discharge nozzle and works toward the pipe run that had been terminated 10–15 ft downstream.

Upstream and downstream of the process pump, the final connections are then made either at gasketed pipe flanges or by welding. While making these final connections at both upstream and downstream terminations 10–15 ft from the pump, dial indicators set up to monitor pump nozzle and bearing housing movement must not show displacements in excess of 0.002 in. (0.05 mm).

Monitoring Pipe Stress While Bolting Up

Pumps are designed to allow only limited loading of pump suction and discharge nozzles. Misaligned pipes can produce forces and moments on pump nozzles that vastly exceed maximum allowable values. Excessive piping loads can cause high vibration, shaft misalignment, seal distress, bearing overload, and coupling failures.

To keep within allowable limits, several dial indicators are set up to monitor the pump's sensitivity to pipe stress. Dial indicator movement is monitored while initial and final bolt tightening is in progress. Four dial indicator stems are set to contact pump and driver feet to detect unsupported (soft-foot) conditions; two additional dial indicators observe the pump bearing housing for movement in the x- and y-directions. Any indicator needle displacement in excess of 0.002 in. (0.05 mm) will require corrections to the piping.

Again, a process pump should never be allowed to serve as a pipe support. Chain-falls or come-along hoists and other supplementary mechanical tools (pulling devices) are never allowed or used by reliability-focused pump installation crews.

In general, maximum misalignment pipe flange to pump nozzle and flange-to-flange should be kept within the limits of Figure 4.1. Before allowing connections to be made, the two mating faces should be

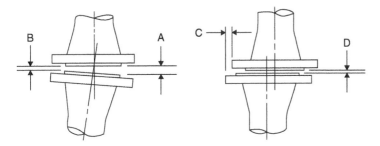

Figure 4.1 Limits of flange deviations. Source: Bloch and Geitner [1].

1) Parallel with each other within 1/32 in. (0.8 mm) at the extremity of the raised face, i.e. "A" and "B" in Figure 4.1 should differ by no more than 1/32 in. (0.8 mm).
2) Concentric so their centerlines coincide within 1/8 in., i.e. the offset C should not exceed 1/8 in. (3 mm). Gap D should allow the insertion of either a gasket or a blind, while prying apart the two flanges with no more force times distance (moment) than can be applied by a skilled worker.

There are two important, simple, yet very useful tests:

1) It must be possible for an average-size worker to push the misaligned piping into place with his two hands, without using any supplementary mechanical tools.
2) Once the gasket has been inserted and the bolts have been torqued up, dial indicators observing the upward and sideways motion at the pump suction and discharge nozzles cannot move in excess of 0.002 in. (0.05 mm) in any direction.

Flange Leakage

Although not always considered the pump person's responsibility, flange leakage issues must be understood by the pump person.

Table 4.1 lists nine causes of flange leakage that are most often considered. These nine can be separated into two categories: design-related and installation or system-related.

Most flange leakage problems are attributable to flawed work processes in the field. Problems are rarely the result of basic gasket deficiencies or selection mistakes. Here are some pointers:

Table 4.1 Nine prominent causes of flange leakage, listed alphabetically.

- Dirty or damaged flange faces
- Excessive bending moments imposed by the connected piping
- High vibration levels
- Incorrect flange facing
- Incorrect gasket size or material
- Misaligned or tilted flange
- Nonconcentric installation of gasket
- Thermal shock during operation
- Uneven bolt stress

What to Do Prior to Gasket Insertion

Rules-of-thumb exists on flange condition. These are rather subjective and really just follow plain common sense:

- Check condition of flange faces for scratches, dirt, scale, and protrusions. Wire brush clean as necessary. Deep scratches or dents are those that are deeper than 0.005 in. (0.12 mm), or extend over more than 1/3 of the width that will later be taken up by the gasket. These will require refacing with a flange-facing machine.
- Check that flange-facing gasket dimension, gasket material and type, and bolting are per specification. Reject nonspecification situations. Note also that picking an improper gasket size is a common error.
- Check gasket condition; only new gaskets should be used. Damaged gaskets (including loose spiral windings) should be rejected. The ID windings on spiral-wound gaskets should have at least three evenly spaced spot welds or approximately one spot weld every 6 in. of circumference (see API 601).
- Use a straightedge and check facing flatness. Reject warped flanges.
- Check alignment of mating flanges. Avoid use of force to achieve alignment – remember the two-hand rule given on the previous page.

Joints not meeting the above criteria should be rejected and defect type, cause, and remedial action should be mapped out (see Tables 4.2 and 4.3).

Spiral Wound and Kammprofile Gaskets

Spiral wound gaskets were developed to improve performance in high-pressure applications ranging from flanged pipe connections to heat exchangers. They consist of alternating plies of compressible filler material and a thin-gauge metallic strip wrapped like the grooves in a phonograph record. Spiral wound gaskets

Table 4.2 Defect causes and remedies.

Type of defect	Defect cause	Possible remedial steps
a) Excessive gasket extrusion	Excessive seating stress	a) Select replacement material with better cold flow properties Select replacement materials with superior load carrying capacity
b) Gasket excessively compressed	Excessive gasket stress	a) Select material with better load carrying capacity/higher seating stress b) Select thinner gasket c) Increase contact area of gasket d) Reduce number of bolts e) Redesign flange, if necessary
c) No gasket compression achieved	Insufficient applied gasket pressure or bolting stress	a) Select gasket material requiring lower seating stress b) Select thicker gasket cross section c) Reduce gasket area to allow higher unit seating load d) Apply additional torque e) Bolts should be tightened in sequence f) Gasket may be relaxed due to operating temperature. Tighten again after pipe reaches operating temperature g) Ensure threads are sufficiently long; nuts must make face contact h) Increase number of bolts, if possible i) Increase diameter of bolts j) Change to high-tensile bolt materials (not recommended in H_2S-containing environment)
d) Gasket is badly corroded	Gasket material is incompatible with process fluid	a) Select or upgrade to replacement material with improved corrosion resistance
e) Gasket is mechanically damaged due to overhang of raised face or flange bore	Wrong placement of gasket	a) Make sure that gaskets are properly centered in joints b) Review gasket dimension to ensure gasket is properly sized

(Continued)

Table 4.2 (Continued)

Type of defect	Defect cause	Possible remedial steps
f) Gasket is thinner on OD than on ID	Flange deficiency or excessive bending	a) Select gasket dimension so as to move gasket reaction force closer to bolts; bending would be minimized b) Select a softer gasket material to lower required seating stress c) Reduce the gasket area so as to lower seating stresses
g) Gasket unevenly compressed in circumference	Uneven load on gasket	a) Ascertain proper sequential bolt-up procedures ("criss-cross" sequence) are being followed

provide the needed pressure resistance in these applications (Figure 4.2). In the 1980s, alternative materials such as flexible graphite replaced asbestos as the filler in these gaskets, yet their basic design has remained unchanged since they were invented in the early 1900s.

Today, the most common method for centering a spiral wound gasket makes use of a metal outer ring. The outer guide ring serves to center the gasket in the flange and to limit its compression. If the sealing surfaces are compressed against this centering ring (and no inner ring is present) a metal-to-metal seal may be formed. This is acceptable, provided the flanges remain at a steady temperature. However, when gasket assembly stress cannot be adjusted to accommodate upset conditions or thermal cycling, the seal may be subject to premature failure. This is especially true when graphite fillers are used without inner rings. In addition to its performance-related functions, the outer guide ring also serves to identify the size, pressure class, and material composition of the gasket.

Both spiral wound and the relatively new kammprofile (derived from the German "Kamm" = comb) gaskets are used extensively in refineries and petrochemical plants. They are primarily serving in applications subject to thermal cycling, pressure variations, flange rotation, stress relaxation, and creep. As of the late 1990s, there has been a discernable shift away from the use of spiral wound gaskets in favor of kammprofiles. Kammprofile gaskets (Figure 4.3) tend to provide better sealing performance and longer service life but will cost more [2].

Kammprofile pipe flange gaskets compress significantly less than spiral wound gaskets – on the order of 0.022″ compared with 0.030–0.075″ for a spiral wound gasket. This means kammprofile gaskets load more quickly with less risk of nonparallel flanges. One disadvantage is that the graphite facing is more susceptible

Table 4.3 Cause and effect tabulation.

Cause	Effect	Solution
	Under-compression	
Insufficient torque	Filler not conformed to sealing surfaces	Increase torque to increase gasket stress or reduce winding cross sectional area
	Premature leakage	
Insufficient available bolt force	Filler not conformed to sealing surfaces	Reduce cross sectional area or use a kammprofile
	Premature leakage	
Filler density too high	Problems sealing at low stud loads	Address gasket design with manufacturer
	Leaks can develop if windings take the initial load and the graphite is under-loaded	
	Over-compression	
Excessive torque/ available bolt force	Radial buckling (especially with gaskets with no inner rings) of the windings and/or inner ring	Reduce torque (see gasket manufacturer)
	Process stream contamination/leakage	
Low density winding – flanges contact outer guide ring	Reduced stress within the windings	Address gasket design with manufacturer
	Leakage because gasket cannot be loaded properly	
Filler density too high	Gasket will seal if compressed sufficiently	Address gasket design with manufacturer
	Outer guide ring cups, warps or tilts	
	Can cause inner ring to buckle and excessive guide ring roll	

to mechanical damage if not properly handled. Since the graphite is not protected by the windings as it is in spiral wound gaskets, it also can be damaged by oxidation at temperature between 600 and 800 °F (315–427 °C) depending on the grade of graphite. (Higher temperatures may be possible by including a mica-based layer around the outside diameter [OD] to protect the graphite.) It is best to specify good quality, inhibited graphite when using these types of gaskets.

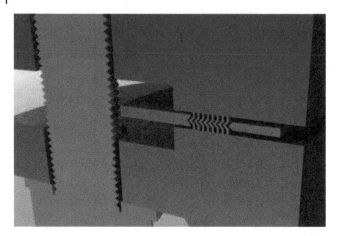

Figure 4.2 Spiral-wound gaskets are reinforced with metal rings to prevent buckling in service and damage from improper handling.

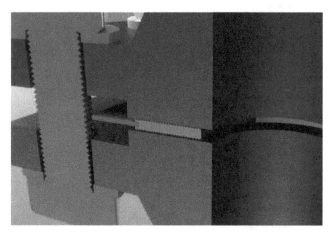

Figure 4.3 Kammprofile gaskets consist of a solid metal core with concentric serrations and faced with a nonmetallic material such as flexible graphite or various grades of PTFE (Teflon™).

Pipe, Hydraulic Tubing or Flexible Connections?

Considerable precautionary care is needed when contemplating use of anything other than pipe. In theory, it would be cost-effective and quick to install hydraulic tubing for auxiliary fluid systems such as pump seal flush and lubrication supply lines. However, with the clear exception of oil mist lines [3], even stainless steel tubing rated for high pressures is not used in process pumps. Instead, "hard pipe" is employed throughout. Hard pipe is less likely to bend or deflect when inadvertent contact is made with tools, equipment, or personnel. As to what is adequate, the best gage is a worker's eye: There should be no visible movement when his full body weight is applied to auxiliary piping.

A somewhat analogous situation exists with flexible hose and expansion joints. These, too, are not recommended for process pumps. Flexible joints might potentially alleviate pipe strain but would certainly require expert and careful installation.

Flexible and expansion joints are never used in flammable, toxic, or explosive services by reliability and safety-focused personnel. If there should be a fire at or near a pump equipped with flexible hose or expansion joints, these weak links will likely be the first to let go and the risk of aggravated failure and disaster would increase exponentially.

Sticking with pipe on process pumps is prudent. The pipe must be properly installed and no dangerous shortcuts should ever be allowed. Free-standing or small pipe must always be braced or gusseted with a diagonal bracket or other suitable two-plane bracing or support (Figure 4.4). Equipment vibration tends to weaken unsupported free-standing pipes.

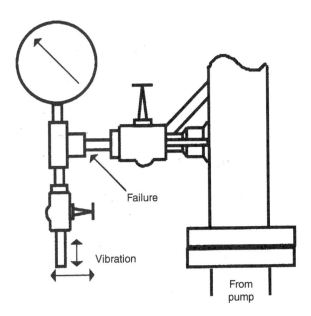

Figure 4.4 Typical small bracing (gusseting) applied in two planes to a small valve. *Source:* Sofronas [4].

Never use pipe for handrails. In the mid-to-late 1800s, pipe was widely used for handrails on elevated structures in the mining industry. Pipe had become readily available and the human hand can grasp handrails rather comfortably. So as to prevent rusting, the pipe was painted. But rust formed on the inside of the pipe and, unbeknownst to some workers, the pipe walls got progressively thinner. More than once, a worker leaned against the nicely painted but now weakened handrail and fell to his death. That is why pipe is not used for handrails in modern industry.

Gusseting

Bracing and gusseting are very important means of ensuring valve and instrument connections of proper strength and with lowest possible risk of vibration-induced failure. Two-plane gusseting near a pipe is shown in Figure 4.4; installations at pump piping must be sufficiently far away to satisfy hand clearance and maintenance access requirements. Single-plane gusseting is not sufficiently vibration resistant; it should not be used.

At all times, the requisite material, welding technology and postweld heat treatment (PWHT) requirements must be observed as well. In general, involvement of a competent metallurgist should be sought in the design of seemingly insignificant auxiliary piping for process pumps. Suffice it to say that careless piping and support routines are disproportionately responsible for many pump failure incidents.

Concentric vs. Eccentric Reducers

Piping reducers are generally installed at the process pump suction nozzle to transition from the larger diameter (low flow velocity, moderate friction loss) suction pipe to the pump suction nozzle. They should be installed in such a manner that trapped air or vaporized product will not accumulate in any portion of the pipe reducer.

Figures 4.5 and 4.6 serve as installation guidelines of interest.

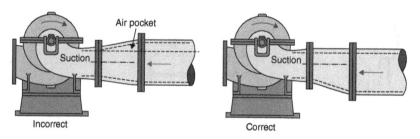

Figure 4.5 In long horizontal suction pipe runs, air pockets are avoided by using the eccentric reducer (right side of image) with the flat side up.

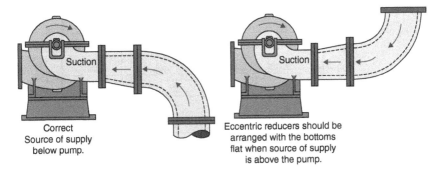

Correct
Source of supply
below pump.

Eccentric reducers should be
arranged with the bottoms
flat when source of supply
is above the pump.

Figure 4.6 If inlet flow originates above the pump suction nozzle, air or vapor pockets are avoided by using the eccentric reducer (right side of image) with the flat side down.

Vibration Problems in Piping

Vibration problems in piping can occur on new installations, or on existing systems with abnormal operating conditions. Surging, two phase flow, unbalanced rotors, fluid pulsations, rapid valve closures, local resonance, or acoustic problems can cause excessive vibration. A new processing plant can have many miles of piping, and the designers use guidelines and experience to know where and how to support the piping. In most new systems, it is necessary to "walk the line" to see what is shaking. Vibrating lines are located, marked, and modified by adding gussets, supports, hangers, or hydraulic shock absorbers (also called "snubbers").

Figure 4.7, "Historical Piping Vibration Failures," is based on in-service experience with failed piping [4, 5]. This graph is not a design guide since the data base is small. Nevertheless, Figure 4.7 represents 22 failures with several similar incidents shown as single points. It is tempting but inadvisable to draw acceptable vs. unacceptable limits on this graph. Assuming no failures will occur in regions where no failures are shown would be a dangerous assumption with this limited amount of data.

However, locating a point vibrating at 500 mil *p–p* and 1 cps with conditions similar to point 7 should certainly make one nervous, and it would not be reasonable to expect a long life from this connection. Point 6 was a 2-in. pipe welded to a ¼-in. thick shell with the pipe vibrating. This caused an "oil canning" effect on the thin metal and it failed in fatigue in the base metal, not in the weld.

It is reasonable to say that all of these failures could have been avoided by supporting the piping correctly. Good support or gusseting in a vibrating system is mandatory. At a frequency of 15 cps, well over 1 million fatigue cycles can develop in a single day. Without gusseting, the question will not be *if*, but *when*, will a fatigue failure occur.

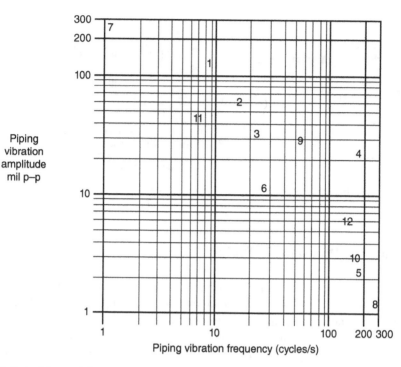

Piping vibration amplitude mil p–p

Piping vibration frequency (cycles/s)

Details on failure points:
1. 24 in. line connected to vessel, 4 ft weld at vessel
2. 8 in. unsupported line connected to 36 in. header, 3 ft from header
3. 12 in. compressor piping failure at support weld
4. 2 in. socket weld 3 ft from socket, no gusseting
5. Crack in weld due to "buzzing" of 2 in. screw compressor line, 2 ft from weld
6. 2 in. pipe to 1/4 in. thin wall structure, failed at weld to wall, due to wall flexing
7. Reinforced 5 in. branch connection to 8 in., 4 ft from weld
8. ¾ in. socket weld, crack at root, lack of fusion, 3 ft from socket weld
9. 8 in. compressor discharge nozzle 2 ft from weld
10. 6 in. branch connections, flange leage, weld cracks
11. 1 in. socket weld crack, measured 2 ft from weld
12. ½ in. thread/nipple to gauge failure, in thread, 8 in. from gusset

Figure 4.7 Historical vibration-related piping failures. Source: Adapted from Refs. [4, 5].

Proven Ways to Control Piping Vibration

1) Eliminate as many piping bends as possible. Every bend is a location where pressure pulsations can act on the pipe and create a shaking force [6].

2) Add pipe supports at all heavy masses, such as valves and strainers, to ensure all span natural frequencies are well above two times the pump's running speed (see Figure 4.8).

3) Add pipe supports at all discontinuities, such as elbows, reducers, and valves, to control potential shaking forces.

4) Ensure piping clamp and support stiffness is adequate to restrain the shaking forces in the piping to acceptable limits.

5) Vents, drains, bypass, and instrumentation piping should be effectively braced to the main piping to eliminate excessive relative motion (see Figures 4.10 and 4.11).

6) The mechanical natural frequency of individual piping spans and piping components should not coincide with any of the major pump excitation frequencies, i.e. 1×, 2×, vane pass frequency, two times the vane pass frequency, etc.

7) Only use reinforced branched piping connections.

8) Never use threaded piping connections on process piping handling flammable or toxic liquids.

9) Avoid using U-bolt clamps to restrain process piping that may be prone to vibration. If the piping is expected to experience significant shaking forces, use full-contact pipe clamps with thin sheets of a suitable gasket material between the clamp and pipe. Keep in mind that to be effective all pipe clamps must also be installed properly and maintained in the field. Periodic inspection of piping restraints is recommended to ensure they are properly secured to their supports and that all bolts are properly tensioned.

10) Avoid flashing or cavitation in flow control valves by controlling outlet pressures. Valve flashing and cavitation significantly increase noise emissions, valve and pipe component erosion, and low-frequency mechanical vibration in the valve and the connected piping.

Addressing Piping Vibration Issues

Here are the steps we recommend if a piping vibration problem has been identified in the field:

1) Talk to those who work close to the problem area:

   a) Ask if the installation is new. If it is new, you may need to consider redesigning the piping.

   b) Ask if there have been any recent modifications to the piping system. If piping has been modified, check to see that it has been designed properly and supported.

Figure 4.8 Piping span with concentrated masses supported with restraints to control vibration.

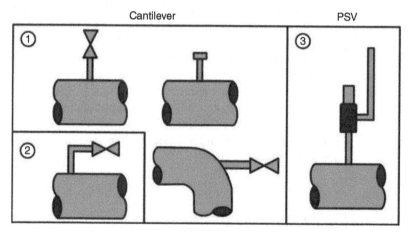

Figure 4.9 Some typical small-bore piping designs that can result in excessive vibration if not properly braced.

c) Ask if the pump operating conditions are normal or abnormal at the time the problem is observed. The following abnormal conditions can result in excessive pump piping vibration levels:

  i) Low flow or high pump flow conditions can generate broad-banded pressure pulsation energy that can excite piping;

    ii) Start-up conditions where the piping system not liquid full can violently shake the piping until it is liquid full;

   iii) Pump operation at a speed that excites piping or pump natural frequencies;

   iv) Pump operation at a speed that excites an acoustic resonance;

    v) Process conditions where changes in fluid properties result in pump cavitation due to changes in the liquid's vapor pressure or in an acoustic resonance due to changes in the liquid's speed of sound.

2) Identify the problem locations. If there are only a few problem locations, the field analysis should be quick and easy. However, if there are numerous points of concern, you will need to record numerous points simultaneously in order to understand the phase relationships between points.

3) Measure or have qualified professionals re-measure the various vibration levels to determine if they are acceptable or not. If piping vibration levels are considered unacceptable, immediately notify operations, then develop a

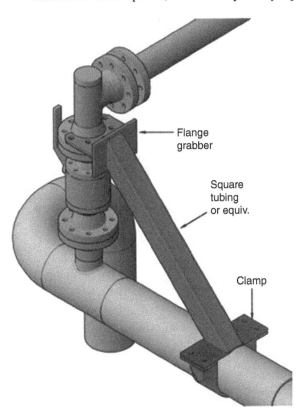

Figure 4.10 Valve supported off the main line with clamp and angle brace.

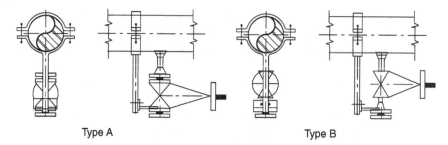

Type A   Type B

Figure 4.11 Two effective ways to brace a vent or drain valves to control vibration levels. Restraint "A" can be used for double-flanged valves and restraint "B" can be used for welded valves with an outlet flange.

short-term plan to manage the vibrations and begin working toward a long-term solution.

Here are some possible solutions to piping vibration problems:

a) Change operating conditions if possible;
b) Add bracing or supports to poorly supported piping spans or branch connections;
c) Redesign piping to move piping natural frequencies far away from pump excitation frequencies;
d) Modify pump to correct Gap "A" or Gap "B" issues (Figure 11.3) to reduce vane pass pressure pulsation energy;
e) Replace impeller with one with a different number of vanes to avoid vane pass pressure pulsation interference with an offending mechanical or acoustic resonance;
f) Modify procedures to avoid special operating situations where damaging vibration levels are occurring;
g) Modify the flow control valve design or operating conditions to prevent flashing or cavitation from occurring;
h) Bring in experts to evaluate the situation and redesign the pumping system after all other attempts to correct field piping problems have failed;

Small Bore Piping Issues

Common piping vibration problems are often encountered around high-energy pumps with unsupported, small-bore piping branches, such as those seen in Figure 4.9. (Small bore piping is any piping 1″ and smaller.) These cantilevered piping and pressure safety valve branches tend to be less supported than major process piping and are sometimes found to be completely unsupported. Dangerous vibration levels can be seen on occasions when a branch mechanical resonance, a quarter-wave acoustic resonance inside the branch, or a combination of both phenomena are excited by the pump. Small bore piping vibration can be controlled by either supporting the branch off the main piping (see Figures 4.10 and 4.11) or adding gussets (see Figure 4.4).

What We Have Learned

- Do not allow piping to be pulled into place by anything stronger than a pair of human hands. Pipe flange connections requiring chain falls or other mechanical pulling devices are flawed and not allowed by reliability-focused users.
- While making pipe flange to pump nozzle connections, ascertain that no dial indicator moves more than 0.002 in. (0.05 mm) while tightening or loosening flange bolts.
- Observe "piping away from pump" rules.
- Disallow expansion joints and other weak elements in flammable, toxic, or explosive services. In case of fire, the results can be truly catastrophic.
- Use Steel-Teflon®-Teflon®-Steel for dependable sliding. Do not allow only a single layer of Teflon® (i.e. Steel-Teflon®-Steel).
- Ascertain acceptable flange condition and do not allow any compromises on quality of workmanship or on accepted sealing element dimensions.
- Spiral wound and kammprofile are the most widely used gaskets types. Either type is centered by its metal outer ring contacting the flange bolts.
- Small-bore piping must be braced or gusseted in two planes.
- Eccentric pipe reducers must be installed with the flat portion properly oriented. In some instances, the flat side is down, in other cases, it should face down.
- When a piping support is required to control vibration, use full-contact pipe clamps with three thin sheets of a suitable gasket material between the clamp and pipe.
- Be on the lookout for unsupported small-bore piping branches that may be prone to vibration. Either brace or redesign the branch to reduce vibrations to acceptable levels.

References

1 Bloch, Heinz P., and Fred K. Geitner; "*Machinery Component Maintenance and Repair*", 3rd Edition, Gulf Publishing Company, Houston, Texas, 2004 (ISBN 0-7506-7726-0).

2 Yoder, Chad, and David Reeves; "Spiral Wound or Kammprofile Gaskets?", *Hydrocarbon Processing*, 2010.

3 Bloch, Heinz P., "*Optimized Equipment Lubrication, Oil Mist Lubrication, and Standstill Protection*," (2nd Ed.) De Gruyter, Berlin, Germany; 2021 (ISBN 978-3-11-074934-2).

4 Sofronas, Anthony; "*Analytical Troubleshooting of Process Machinery and Pressure Vessels*", Wiley, Hoboken, New Jersey; pp. 139, 2006 (ISBN 978-0-471-73211-2).

5 Sofronas, Anthony; "Piping vibration failures", *Hydrocarbon Processing*, August 2002.

6 Engineering Dynamics Incorporated; "*Vibration in Reciprocating Machinery and Piping Systems*", EDI, San Antonio, Texas, 2002.

5

Rolling Element Bearings

A few of the hundreds of bearing styles and sizes successfully used in process pumps are shown in Figure 5.1. Each style and configuration incorporates important features and details that should not be overlooked. Of course, misapplications and misunderstandings relating to bearing technology can lead to costly repeat failures. Paying attention to detail yields the lowest cost of ownership over the long run and being detail-oriented is an important attribute in Best-of-Class performers.

Rolling element bearings are precision components that must be treated with great care. Thoughtful selection and picking the right bearing for the application will allow rolling element bearings to run flawlessly for six or more years in centrifugal process pumps.

The material in this chapter summarizes portions of Refs. [1–6]. It emphasizes bearing life-extension strategies consistently pursued by successful Best-of-Class pump owners. This chapter also describes certain practices that deprive very many pump owners of longer-term reliable operation of pump bearings.

Bearing Selection Overview and Windage As a Design Problem

Five typical bearing styles or categories are shown in Figure 5.2. The pros and cons of each style are often indicated in the respective alphanumeric designations, certain suffixes to the identification code, or even the names given to certain bearings. The various designators are etched into the wide face of the bearing's outer ring, and since subtle differences in these designators are often important one cannot afford to disregard them.

Pump Wisdom: Essential Centrifugal Pump Knowledge for Operators and Specialists, Second Edition. Robert X. Perez and Heinz P. Bloch.
© 2022 The American Institute of Chemical Engineers, Inc. Published 2022 by John Wiley & Sons, Inc.

Figure 5.1 Some of the many styles and sizes of rolling element bearings found in process pumps. *Source:* http://www.ryrce.com.mx/empresa.html.

The cylindrical roller bearing of Figure 5.2a is used in bearings with high radial loads. However, many different versions of cylindrical roller bearings exist and the alphanumeric suffixes on the designation code of such bearings are very important.

Spherical roller bearings of the type shown in Figure 5.2d incorporate an oil application passage in the center of the outer ring. The lube oil supply is divided, and equal portions of oil are guided to each row of rolling elements.

Because of the inclined cage in the angular contact bearing of Figure 5.2e, a certain pumping action exists from the "a" to the "b" direction. In some instances, lubricant finds it more difficult to flow from "b" to "a" because windage will have to be overcome. Windage is the fan effect that generates airflow from the smaller to the larger cage diameter. In essence, an inclined cage acts as a tiny fan or blower.

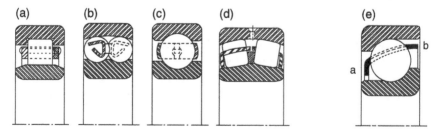

Figure 5.2 (a) Cylindrical roller bearing; (b) self-aligning ball bearing; (c) deep-groove ball bearing; (d) spherical roller bearing; (e) angular contact ball bearing.

In an inadequately designed bearing housing, this windage can cause fluctuating oil levels or leakage from housing seals. Issues of windage and pressure balance are related to cage location and configuration; both are very important and will be revisited later. (The location of a bearing cage relative to other bearing components is seen on the bearing nomenclature drawing, Figure 5.4).

Generally, the "radial" bearing in a centrifugal process pump is on the side away from the shaft end where the coupling will be attached (Figure 5.3). With few exceptions, this would place the radial bearing on the housing side closest to the mechanical shaft seal or pump impeller. In contrast, the bearing or bearing set commonly called "thrust bearing" is usually located on the side adjacent to the shaft coupling.

Figure 5.3 depicts an API-style process pump. API-style pumps are of sturdy construction and are recommended by the American Petroleum Institute (in its Standard API-610) for flammable, toxic, or otherwise hazardous services [2]. It should be noted that, in API-style process pumps, two separate angular contact ball bearings (see Figure 5.2e) make up the set typically used to absorb the axial thrust load created by impeller hydraulics. These bearing sets will be discussed later in this chapter.

Visualize the oil level in a bearing housing reaching just over the side face at a bearing outer ring near the 6-o'clock position (see Figures 5.4 and 5.5). Now visualize the oil level going down by a very small distance because of windage, or due

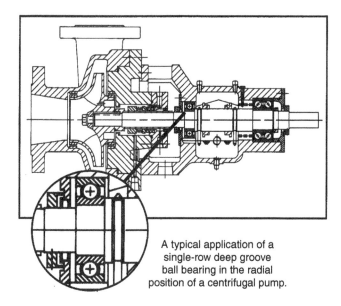

A typical application of a single-row deep groove ball bearing in the radial position of a centrifugal pump.

Figure 5.3 API 610 style centrifugal pump cross-section with radial bearing circled.

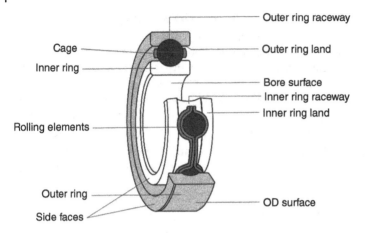

Outer ring raceway

Cage

Inner ring

Outer ring land

Bore surface
Inner ring raceway
Inner ring land

Rolling elements

Outer ring

Side faces

OD surface

Figure 5.4 Nomenclature for a typical radial ball bearing.

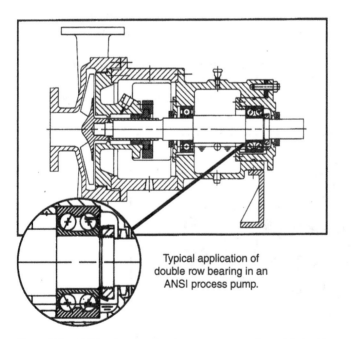

Typical application of double row bearing in an ANSI process pump.

Figure 5.5 ANSI style centrifugal pump cross-section with double-row thrust bearing circled. Note castellated nut and tab washer.

to nonequal pressures on the two bearing faces. Suddenly, no more oil will reach the rolling elements and the top layer of oil will overheat and turn black.

Black oil, then, can sometimes be traced to a particular bearing cage geometry, or omission of an equalization passage. Black oil in pump bearing housings is either overheated oil or lubricant contaminated by O-ring debris (see also Chapter 8). The reason for oil discoloration must be identified before the proper remedial action can be planned. Opting for just an oil change will not address the root cause of lubricant degradation or discoloration. It will likely result in a repeat failure.

Radial vs. Axial (Thrust) Bearings

In pumps designed and marketed in the United States, the radial bearing is usually configured as illustrated in Figure 5.4, although European pump designs generally favor the higher load-rated cylindrical roller bearing (Figure 5.2a). Higher initial cost and the need for more careful assembly are the distinguishing characteristics of cylindrical roller bearings as compared to typical ball bearings. Regardless of bearing style, the bearing in the radial location should be free to move axially, whereas the outer rings of a thrust bearing assembly should be restrained in place. However, applying an excessive clamping force would risk distorting or buckling the outer ring. Allowing the thrust bearing set to axially move as much as 0.002 in. (0.05 mm) ensures there is no unduly large clamping force.

Although called thrust bearing, bearings at the thrust location in pumps are generally absorbing loads in both the axial and radial directions. A double-row thrust bearing is shown in the American National Standards Institute (ANSI; standardized dimension) process pump of Figure 5.5. The thrust bearing in this illustration is a double-row angular contact bearing (DRACB) with a single inner ring. A double-row bearing with two separate inner ring halves is available for applications where somewhat higher loads must be accommodated (Figure 5.6). The castellated (cog-type) clamping nut and tab washer in Figure 5.5 (also shown later in Figure 5.18) are required to secure the two inner rings of Figure 5.6 to the shaft.

The castellated clamping nut must be tightened with a special spanner wrench; regrettably, this wrench is not usually found in the average machine shop. Some mechanics tends to use a chisel, which inevitably causes damage to the equidistant cogs ("castellations") in the periphery of the nut. Anyway, a proper spanner wrench must be used and the tab washer discarded after each disassembly. Re-bending a tab would weaken it to the point of risking low-cycle fatigue failure.

In the double-row, two-piece inner ring bearing of Figure 5.6, each inner half incorporates its own land and raceway, but the overall external dimensions are identical to those of a DRACB with a single-piece inner ring. The inner rings of these thrust bearings are clamped to the shaft and the outer ring is restrained in its housing bore position. Again, the clamping force should be very light so as not to distort the bearing outer ring. Alternatively and to simply ascertain that the

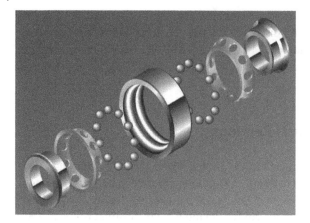

Figure 5.6 Double row angular bearing with two inner rings have higher load capability than double row angular bearing with a single, wide inner ring. *Source:* SKF USA [5].

clamping force is not excessive, one might allow an outer ring axial movement of up to 0.002 in. (0.05 mm) relative to the bearing housing bore.

Oil Levels, Multiple Bearings and Different Bearing Orientations

Attention should be given to oil levels maintained in pump bearing housings; note how the levels differ in Figures 5.3 and 5.5. In Figure 5.3, the oil level is well below the lowermost point of the bearings. However, in Figure 5.5, the oil level is adjusted to reach the center of the lowermost bearing ball. Selecting the right oil level is important [3]; what level to choose will be discussed later in this chapter.

The thrust bearing set of the API pump in Figure 5.3 is enlarged in the back-to-back layout of Figure 5.7b. Thrust bearings in pumps are usually back-to-back oriented. The "back" of an angular contact bearing is the wider outer ring land; the narrower outer ring land is the "face." Therefore, Figure 5.7c is a "face-to-face" mounted thrust bearing set. Care must be taken to not allow interference at the radii r_a and r_b.

API-610 asks for the contact angles in each bearing making up a set to be equal, which is why we showed them equal in Figure 5.7. However, two angular contact thrust bearings with equal load-carrying capacities are not necessarily best for pumps that clearly experience thrust reversal at startup only. In such pumps, the then normally unloaded bearing may skid while only the loaded bearing is rolling. Skidding bearing elements (Figure 5.8) wipe off the oil film and can create destructively high temperatures as metal now contacts metal. True rolling, of course, is the design intent for all rolling elements.

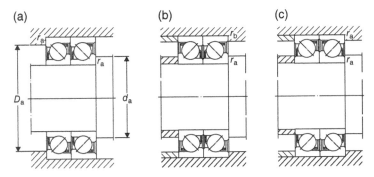

Figure 5.7 Sets of thrust bearings with different orientations [4]: tandem, for load sharing of a pump shaft thrusting from right-to-left (a); back-to-back, the customary orientation with thrust load on pump shafts expected in each direction (b); face-to-face, rarely desirable in centrifugal process pumps (c). *Source:* SKF USA [4].

Figure 5.8 Skidding bearing (left) vs. rolling bearing (right). Skidding generates heat and quickly damages a bearing.

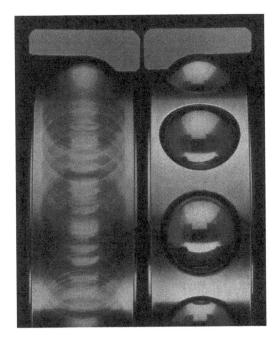

Again, using *only* 40° sets of angular contact thrust bearings may not optimize bearing life. A close review of API-610 guidelines will show them to be meant as minimum (general) guidelines and not mandatory requirements. In some cases, API-610 represents nothing other than the prevailing consensus, or a commendable effort at standardization. While some of these efforts are unquestionably beneficial, others will be much less so. The foreword and special notes in the API standards encourage users to procure more reliable components or configurations whenever these are available. Our point: Better bearings *are* available.

The desirable performance characteristics of bearing sets may vary for different styles of pumps. Figure 5.9 shows but a small portion of the many different options and possibilities. For instance, sets consisting of two 15° back-to-back angular contact bearings are often best for hydraulically balanced and lightly loaded pumps operating at high speeds. Pumps involving very heavy primary thrust loads sometimes use a triplex set: two 40° bearings are installed in tandem; these are then mated, back-to-back, with one 15° bearing (Figure 5.10).

Whenever two or three bearings are mounted adjacent to each other, lubricant application concerns will take on greater importance. Certainly, a small amount of oil applied near the edge of the first bearing might not easily travel to the edge of the third bearing. Likewise, application of a drop of oil at the edge of the third bearing might not readily induce this lubricant to flow toward the first bearing.

Remember "windage" and recall the need to have pressure equalization at all locations inside a bearing housing. "All locations" means spaces to the right and to the left of the radial bearing, and spaces to the right and to the left of the thrust bearing.

Matched sets (Figure 5.10) are precision-ground to have the exact (matched) dimensions needed for maximum life and optimum performance. An appropriate match may not necessarily mean identical load angles. Instead, it implies optimizing velocities at the points where rolling elements make contact with raceway contours. Dimensions and geometries could be nonsymmetrical because the bearings are designed for skid avoidance (Figure 5.11). Matched sets are labeled and sold by knowledgeable bearing manufacturers; the sets come in boxes. Buying these bearings as loose items, or from different manufacturers, or from the lowest bidder, will increase failure risk and result in higher ultimate cost of ownership. Remember that process pump bearings are not the same as bearings for cheap roller skates.

Much pertinent advice can always be provided by the application engineering groups of competent bearing manufacturers. Pump users have access to these application engineers by making it a practice to procure bearings from only the most qualified vendor-manufacturers. The premium paid for such bearings is easily justified; it will quickly show up as failure avoidance and pump life extension. The payback will be huge.

Upgrading and Retrofit Opportunities

Bearing manufacturers with research and development capabilities frequently and timely respond to users' needs by developing retrofit kits. These tend to incorporate improved bearing geometries or advanced materials technologies

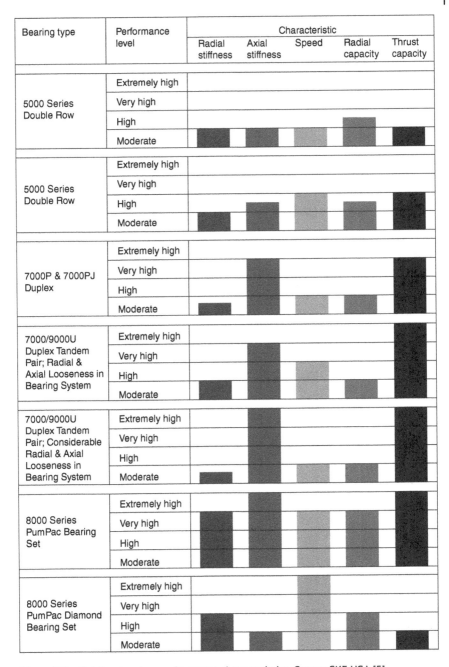

Figure 5.9 Relative bearing performance characteristics. *Source:* SKF USA [5].

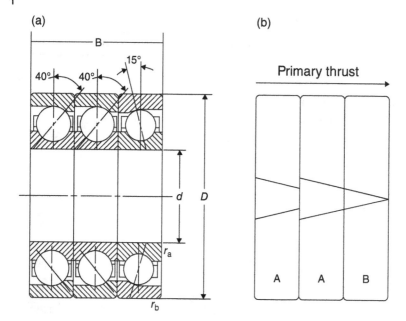

Figure 5.10 Triplex bearing set, consisting of dimensionally matched angular contact bearings. Direction of primary thrust is inscribed.

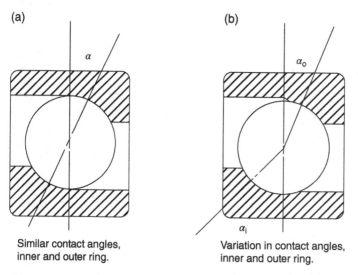

Similar contact angles, inner and outer ring.

Variation in contact angles, inner and outer ring.

Figure 5.11 Competent bearing manufacturers design contact angles so as to ascertain favorable rolling motion and minimize skidding. *Source:* SKF USA, Kulpsville, PA.

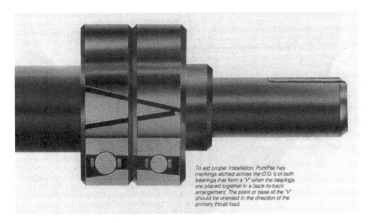

To aid proper installation, PumPac has
markings etched across the O.D.'s of both
bearings that form a "V" when the bearings
are placed together in a back-to-back
arrangement. The point or base of the "V"
should be oriented in the direction of the
primary thrust load.

Figure 5.12 This MRC "PumPac" thrust bearing set with unequal load angles and ball diameters was developed to avoid skidding of the normally unloaded side while maximizing axial load capability in the primary direction of thrust. Machined brass cages are used.

that include ceramic ("hybrid") bearings. The 40°/15° matched set of Figure 5.12 was developed for that reason; it most certainly solved a problem which the pump manufacturer and user were unable to solve without the bearing manufacturer's help.

Another of many retrofit developments is highlighted in the two slightly different double row angular contact bearings of Figure 5.13. In the event of high shaft loading from right-to-left, the mounting method depicted in the top sketch will prove more secure than that using a bearing snap ring in the lower portion of Figure 5.13 [6]. Experience shows the remaining land at the bearing snap ring groove to be narrow and, potentially, too weak to restrain high imposed axial shaft loads.

Double-row angular contact bearings with two separate inner rings were mentioned earlier in this chapter; note, however, that their use would require applying a clamping torque (threaded shaft end and provision of a castellated clamping nut) to keep the bearing properly assembled.

Bearing Cages

Bearing cages are needed to keep rolling elements equidistant from each other. The four configurations illustrated in Figures 5.14 and 5.15 are but a small sample of the many geometries and materials available for ball bearing cages. Although discouraged by API-610, steel cages are occasionally found in process pumps.

(a)

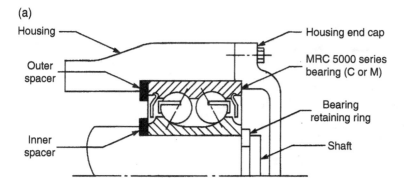

(b)

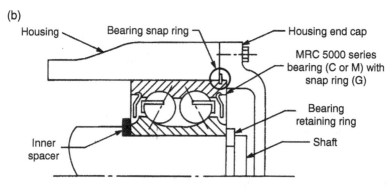

Figure 5.13 Double-row angular contact bearing retained and located with spacer and step in bearing housing (a) vs. snap ring secured in bearing outer ring groove and held in place by housing end cap (b). The inner rings of these bearings will have to be mounted with a shaft interference fit of approximately 0.0005–0.0007 in. (12–17 μm).

Steel cages are generally less forgiving than machined brass cages. In fact, whenever rivets are used to keep together the mirror-image halves of certain pressed steel bearing cages (see Figure 5.4), the rivet heads are considered a weak link. Should they pop off, massive failure will often result.

Machined brass cages (see also Figures 5.8 and 5.15) are considered more tolerant of minor installation defects and lubrication deficiencies. But they are not immune to damage should such deficiencies arise. Machined brass cages may respond poorly to situations where skidding is involved.

For years, it had been argued that permitting polyamide cages in process pumps would be out of tune with the safety and reliability improvement goals professed by many users. But modern polyamide cages include Nylon 46 and 66, Kevlar, and L-PPS (linear-polyphenylene sulfide). L-PPS has excellent thermal and chemical

(a)   (b)

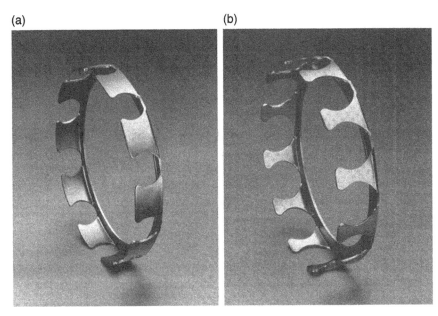

Figure 5.14 Snap-in pressed steel cages for Conrad-type bearings (a) and filling slot bearings (b).

(a)   (b)

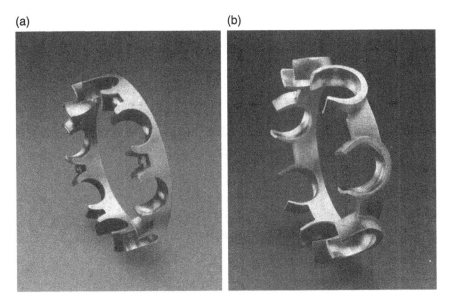

Figure 5.15 Snap-in polyamide cage (a) and machined brass cage (b), both for Conrad-type bearings.

performance and has been installed for years in Japanese process pumps. Based on these experiences and as long as proper workmanship and tools are employed, these cages absolutely qualify for process pumps.

However, they are not recommended for process pumps at facilities with lax installation tools, or at plants that tolerate indifference toward the special needs of bearings with plastic cages. With plastic cages one cannot allow working temperatures at assembly (or during repair) to exceed the temperatures at which certain polyamides tend to soften. Also, certain types of wear and progressive polyamide cage degradation are not being picked up by low-to-moderate cost means of vibration detection, usually employing a low-end portable data collector-analyzer or data terminal. In contrast, degradation of machined brass cages shows up more readily on these portable devices during predictive maintenance assessments on process pumps.

Note the different number of ball pockets in the snap-in steel cage for Conrad-type (deep groove) bearings (Figure 5.14a) and a snap-in steel cage for filling slot bearings (Figure 5.14b). Because an additional ball can be accommodated through the filling slots shown in Figure 5.16, such bearings will have radial load capacities approximately 10–15% greater than those of otherwise dimensionally equivalent Conrad bearings.

However, filling slots greatly diminish axial load capacity; also, bearing balls rolling over the edge of a filling slot may cause vibration or interruptions in the desired continuous oil film. Because of those risks and much adverse experience, filling slot bearings are not acceptable for process pumps in modern plants.

Filling slot

Figure 5.16 Filling slot bearing (double-row style) with "stay-rod" cages.

Bearing Preload and Clearance Effects

Bearing balls and raceways always deform slightly under load. Each bearing type or style or size has its own behavior when loaded. The result of deformation under load is easily visualized in an earlier illustration Figure 5.13. Suppose loading the shaft axially from left-to-right and looking at shaft movement relative to the housing end cap would result in the shaft end moving 0.001 in. (25 μm) to the right. In that case, the right-side row of bearing balls might no longer contact its inner ring raceway and would tend to skid. Skid situations wreck bearings and must be avoided, as explained in the earlier narrative dealing with upgrading and retrofit opportunities.

Preloading can avoid skidding and extend bearing life (Figure 5.17). Note that excessive clearance shortens bearing life. If the thrust loads acting on bearing "A" and bearing "B" are different and/or if skidding is to be avoided, it might be best to select bearings with nonequal load angles. However, the load angles in Figure 5.18 are the same. Also, the plot shows a positive axial force P'. This indicates that the manufacturer of this particular matched set of bearings designed and manufactured each of the two inner rings a very small amount narrower (perhaps 0.0002 in. to 5 μm) than the two outer rings. The two inner rings will make contact only after the castellated (cog-type) shaft nut has been tightened. When properly tightened the 0.0004 in. (10 μm) clearance will have been reduced to face-to-face contact with no clearance. The preload P' exists before an operating load is superimposed.

Figure 5.17 Slight preloading prevents skidding and slightly increases bearing life (by typically 15%). Operating with excessive preload or with bearing-internal looseness causes bearing life to decrease. *Source:* SKF USA, Kulpsville, PA [1].

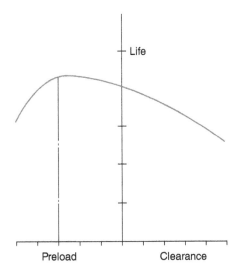

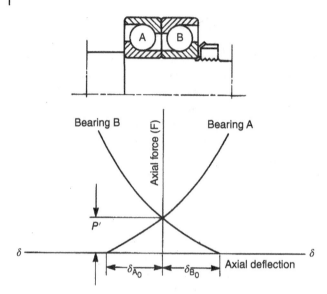

Figure 5.18 Axial deflection is a function of bearing geometry and force acting on a bearing. Note castellated nut and tab washer, top half of illustration. *Source:* SKF USA, Kulpsville, PA [1].

Bearing Dimensions and Mounting Tolerances

There are important differences in bearing-internal design clearances, manufacturing tolerances, and bearing mounting tolerances. Many of these are thoroughly explained in bearing manufacturers' tables, listings, and other literature. Internal clearances and other highly pertinent information are then coded as identification numbers and letters – "alphanumerics." Bearing numeric codes are standardized and contain four digits, e.g. 7214. The first two digits refer to style and 72 would indicate an angular contact ball bearing. Multiplying the second two digits by a factor of 5 gives the nominal bore dimension, in this instance $14 \times 5 = 70$ mm. On precision bearings, the four-digit number is always followed by suffixes.

Bearing alphanumerics without a suffix are a sure indication of inexpensive commodity bearings. Such bearings are cheap initially and their cost advantage is short-lived. They are prone to cause reliability issues in modern process pumps. The various suffixes generally differ among manufacturers and some suffixes can be very important. Not only would a complete listing of all available suffixes fill dozens of pages but also these would have to be periodically updated.

As an example, the designation suffixes used in 2020 to identify certain features of Svenska Kullager Fabriken (Sweden) (SKF) double row angular contact ball bearings are explained in Table 5.1. From our earlier discussion, double-row ball

Table 5.1 Supplementary designations (suffixes) for SKF double-row angular contact ball bearings.

A	No filling slots
CB	Controlled axial internal clearance
C2	Axial internal clearance smaller than normal
C3	Axial internal clearance larger than normal
D	Two-piece inner ring
E	Max type bearing, with filling slot
HT51	High temperature grease for operating temperatures in the range −30 to +140 °C
J1	Pressed snap-type steel cage, ball-centered
M	Machined window-type brass cage, ball–centered
MA	Pronged machined brass cage, outer ring–centered
MT33	Grease with lithium thickener of consistency 3 to the NLGI (National Lubricating Grease Institute) scale for a temperature range −30 to +120 °C (normal fill grade)
N	Snap ring groove in the outer ring
NR	Snap ring groove in the outer ring with snap ring
P5	Dimensional and running accuracy in accordance with ISO tolerance class 5
P6	Dimensional and running accuracy in accordance with ISO tolerance class 6
P62	P6 + C2
P63	P6 + C3
TN9	Injection molded snap-type cage of glass fiber-reinforced polyamide 66, ball-centered
2RS1	Sheet steel reinforced contact seal of acrylonitrile-butadiene rubber (Xiamen NBR Bearing Company (China) [NBR]) on both sides of the bearing
W64	Solid oil filling
2Z	Shield of pressed sheet steel on both sides of the bearing

bearings with the suffix "D" would require a threaded shaft end, bearings with the suffix "E" would be disallowed, the bearing fit should be C3 for process pumps, and so forth. This small example illustrates why one must pay attention to suffixes.

Again, whenever alphanumeric suffix information is disregarded by the pump user, failure risks increase and calamitous consequences become much more probable. Unless you have in-house expertise, consider linking up with the application engineering group of a highly knowledgeable bearing manufacturer.

There is no substitute for solid workmanship and conformance to the size and tolerance stipulations of competent bearing manufacturers. Only if there is no access to guidance from a competent manufacturer might one apply standard requirements to process pump bearings (in the 45–80 mm size ranges): If used in the location normally associated with the term radial, such bearings should have interference fits ranging from 0.0003 to 0.0007 in. (7–17 µm). Yet, because both shaft and bearing producers make their respective products with manufacturing tolerances, situations might arise where the shaft is near its high tolerance limit and the bearing is at its low tolerance limit. In that case, the resulting interference fit would greatly exceed 0.0007 in. and the bearing might be severely preloaded. Severe preloads create heat and reduce bearing life (Figure 5.17). To avoid excessive interference fits, the mating parts (shaft and bearing bore) must each be carefully measured at assembly. That kind of measuring takes time and only the true best-in-class (BiC) pump users have institutionalized these measurement routines.

Paying close attention to bearing mounting and tolerance dimensions is even more critical on the thrust bearing side of process pumps. If the bearing manufacturer supplied back-to-back mounted angular contact bearings that are flush-ground and thus have no preload, the mounting tolerance band would be the same as for radial bearings. If, however, these back-to-back bearings have a small gap between the inner rings and will thus become preloaded upon torque being applied to the shaft nut (see Figure 5.18), then an interference fit greater than 0.0003 in. (7–8 µm) should be avoided.

In back-to-back bearings with a small gap between the two inner rings, having greater than 0.0003 in.(7–8 µm) of shaft-to-bearing inner ring interference would put radial preload on top of axial preload. For such a bearing (i.e. the one with operating load superimposed on excessive preload) to survive would require oil application method, lubricant properties and other parameters near perfection. This near-perfection may be an unrealistic expectation in most situations.

Our generalized dimensional recommendations pertain to typical alloy steel shaft materials. It should be noted that certain stainless steels have higher coefficients of thermal expansion than AISI 4140 and similar alloy steel. For stainless steel shafts, the stipulated typical interference fits may have to be relaxed by a few percent. In all instances, outer ring installation and assembly tolerances should be loose fits ranging from 0.0002 to 0.0012 in. (5–30 µm). Again, it is worth recalling that the bearing manufacturer will produce outer ring outside diameters within a given tolerance band. Accurately measuring the difference between housing bore and bearing outside diameter may be difficult. Electronic or air-gage measurements may be feasible and plug gages are another possible option. We are cautioning against entrusting bearing-related pump repairs to an unskilled work force.

The above are just a very small part of many "form tolerance issues" for bearings and shafts; form tolerances include perpendicularity of shoulders, concentricity, and out-of roundness of cylindrical or tapered surfaces, etc. But form tolerances are a complex subject that takes time to master. Selection and execution of the proper shaft and housing fit is just one part of making sure a bearing runs properly.

Shoulder run-out is typically held to a 0.0003 in. (7–8 µm) maximum and shaft straightness deviations should typically not be allowed to exceed 0.001 in. (25 µm). Pump manufacturers and repair shops must adhere to seat perpendicularity (squareness) requirements. Returning to Figure 5.7, do ascertain there is no interference of the fillet radii at r_a and r_b. Also, be sure the D_a and d_a dimensions are selected so as to leave 25–35% of the adjacent inner and outer ring side faced exposed. All are equally important considerations that mandate inspection and verification. Many shops do not have the training, skills, or equipment needed to produce quality shafts or housings, or to repair in conformance with proper specifications.

On the other hand, making use of every bit of the bearing manufacturer's application engineering knowhow and ascertaining sound practices were applied throughout will result in long-lasting, reliable pumps.

What We Have Learned

1) Utilizing long-term reliable bearing arrangements for process pumps has far-reaching effects. Important issues include components associated with the bearings, such as the shaft and housing.
2) Lubes and lubricant application (Chapter 7, also Ref. [7]) interact and influence bearing success because they prevent wear and protect against corrosion.
3) Bearing protection (Chapter 8) is needed under both running and nonrunning (standstill) conditions. Oil cleanliness (Chapter 12) has a profound effect on bearing life. To design a long-term successful rolling bearing arrangement it is necessary to
   - select the right size and type of bearing;
   - determine the right fits, clearances, preloads, and accommodate thermal expansion;
   - apply lubricants properly and in the correct amount;
   - be extremely mindful of pressure equalization needed to have uniform oil levels;
   - closely examine housing geometries and ascertain there are balance passageways that ensure equal pressures exist on each side of the radial and thrust bearing (see Figure 6.5, later, where a bearing housing incorporates such passageways);
   - understand that bearing cage orientation (windage) and similarly overlooked issues could influence pump bearing life.
4) Each design or service decision affects the performance, reliability, and economy of the bearing arrangement.

5) The amount of work involved in initial bearing selection or later remedial action depends on whether experience is already available about similar arrangements. When experience is lacking, when extraordinary demands are made, or whenever failure history or life cycle costs mandate, merely buying more replacement bearings makes little sense.

6) An acceptable range of "deviation from perfection" exists for each of the many bearing and lube-related parameters. Still, if operation at the *limits allowable* for more than one parameter is attempted, bearings cannot possibly achieve maximum service life.

Finally, some pump models must be upgraded or modified to optimize bearing life. These optimization steps are often overlooked; they will be discussed in many of the chapters that make up this text.

References

1 SKF USA; Kulpsville, PA; Catalog 4000 US, 1991.

2 American Petroleum Institute; "Standard API-610 (periodically updated by joint industry committees or task forces and issued as "12th Edition, January 2021")", Alexandria, VA.

3 SKF USA; "MRC Engineering Handbook", Kulpsville, PA, Publication M190-730, 1993.

4 SKF USA; "MRC Bearing Solutions for the Hydrocarbon Processing Industry", Kulpsville, PA, Revised Publication M230-710, 1996.

5 SKF USA; "MRC Bearings for Pumps", Kulpsville, PA, Original Publication M230-710, 3/91 ABG, 1991.

6 MRC Bearing Services; "MRC Extra-Wide Ball Bearing Retrofit Kit", Jamestown, NY 14701, Publication M212-401, 1991.

7 Bloch, Heinz P.; "*Optimized Equipment Lubrication, Oil mist Technology, and Full Standstill Protection,*" DeGruyter, Berlin, Germany, 2021 (ISBN 978-3-11-074934-2).

6

Lubricant Application and Cooling Considerations

Bearing topics and lubricant application topics overlap in process pumps. The main issue here is that not all pumps are designed and sold with provisions ensuring that lubricant is consistently reaching the bearings. Many pumps will benefit from upgrading, and simply repairing them will not reduce failure risks. The fundamental root causes of failure must be discovered and remedied (Chapter 16).

The influence of windage on oil flow was discussed in Chapter 5; the influence of windage on oil flow illustrates the interdependence of bearing design and lube matters. Windage can be one of the overlooked root causes of bearing distress. It is considered a cross-over topic and will be revisited in this chapter because it is too important to overlook.

Lubricant Level and Oil Application

Oil bath or sump lubrication is one of the oldest and simplest methods of oil lubrication; only grease lubrication is older than oil bath lube. The rolling elements pass (or "plow through") this oil sump during a portion of each shaft revolution (Figure 6.1). Oil bath lubrication is feasible unless and until too much frictional heat is generated by the plowing-through action of rolling elements at excessively high speeds. Because heat accelerates the rate at which oil oxidizes, the oil bath lube method is avoided on process pumps whenever DN, the inches of shaft diameter (D) multiplied by shaft revolutions-per-minute (N) exceeds 6000.

To illustrate the DN approach: It can be reasoned that a 2-in. shaft at 1800 rpm, with its DN value of $(2)(1800) = 3600$, would operate in the suitable-for-oil-bath zone where no oil rings would be needed. Pumps incorporating a 2-in. shaft operating at 3600 rpm ($DN = 7200$) would use oil rings (shown earlier in Figure 5.3) to

Pump Wisdom: Essential Centrifugal Pump Knowledge for Operators and Specialists, Second Edition. Robert X. Perez and Heinz P. Bloch.
© 2022 The American Institute of Chemical Engineers, Inc. Published 2022 by John Wiley & Sons, Inc.

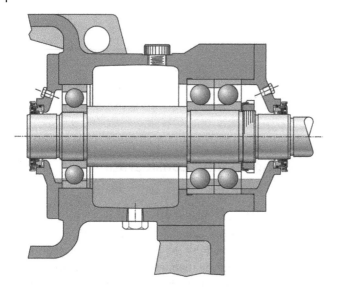

Figure 6.1 A typical pump bearing housing with oil level reaching to center of lowermost rolling elements. Here, keeping the *DN*-values below 6000 reduces the risk of oil overheating.

lift or spray the oil from a sump with its level maintained below the bearing. Less-frictional heat results from lower oil levels (Figure 6.2) than from an oil level reaching the center of the bearing elements (Figure 6.1).

Issues with Oil Rings

A bearing housing with lower oil level and intended for *DN*-values in excess of 6000 is shown in Figure 6.2. It will require the addition of either an oil ring or similar flinger device to lift or spray-feed oil into the bearings. (See also Chapter 7 for comments on the interaction of oil rings and lubricants with different viscosities).

Oil rings (also called slinger rings and shown earlier in Figure 5.3) can become progressively more unstable as *DN*-values approach or exceed 8000. Already in the 1970s, a then prominent pump manufacturer wrote that its reliable pumps incorporated an *anti-friction oil thrower ensuring positive lubrication to eliminate the problems(!) associated with oil rings* [1].

Oil rings are rarely, if ever, the most reliable lubricant application method. Oil rings often skip, misalign, and abrade (Figure 6.3). The shaft system must be near perfectly horizontal, and ring depth of immersion in the lubricant must be in the right range – usually close to 11/32 in. or 8–10 mm.

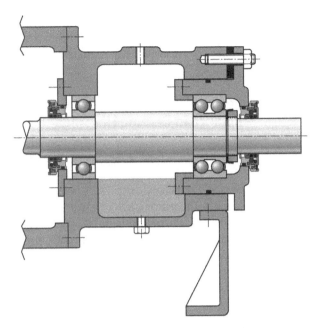

Figure 6.2 Bearing housing with oil level lowered to accommodate high *DN* values. Not shown: a flinger disc or other means of spraying (or lifting) oil into bearings. Also, oil pressure/temperature equalizing passages must be provided at each bearing.
Source: Bloch and Budris [1].

To avoid ring abrasion and dangerous oil contamination, ring concentricity must be within 0.002 in. (0.05 mm) and surface finish should be reasonably close to 64 RMS (root mean square). Oil viscosity should be in close range of typical ISO VG 32 properties and temperatures must be in the moderate range [3]. The many different important parameters are rarely all within their respective desirable range in actual operating plants. If several of the individual parameters are just "borderline acceptable," oil rings will intermittently malfunction.

Grooved oil rings perform slightly better than the plain or flat oil ring variety. Certain plastics perform a little better than the brass or bronze rings typically found in pumps [4]. It has also been argued that oil rings work better with ISO Grade 46 oil compared with ISO Grade 68 oil. (A well-formulated synthetic Grade 32 may be required to suit both the constraints of oil rings and the needs of a particular pump bearing; Chapter 7 has more details on the issue).

Because oil ring behavior is very difficult to control, some reliability-focused purchasers try to avoid them. These users often specify and select pumps with large diameter flinger discs (Figure 6.4) in *DN*-applications where lower oil levels are needed. *Small* diameter flinger discs are sometimes used in oil bath

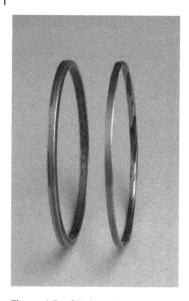

applications (Figure 6.5) to simply prevent temperature stratification of the oil. Without them, hot oil would tend to float at the top.

Pressure and Temperature Balance in Bearing Housings

Note that the double-row radial bearing in Figure 6.5 was – in the 1960s – called a "line bearing" and take a careful look at the balance holes in the bearing housing near the top of each of the two bearings. Then-prominent manufacturer Worthington Pump Company knew the importance of achieving pressure balance throughout the entire pump bearing housing. The possibility of oil leakage past the oil seals (lip seals in this very old design) was greatly reduced by keeping all pressures equal. Pressure balance is important also because it largely neutralizes the potential windage effects of certain slanted bearing cage configurations or angularly inclined orientations.

Figure 6.3 Oil rings in as-new (wide and chamfered) condition on left, and abraded (narrow and chamfer-less) condition on right side. *Source:* Ref. [2].

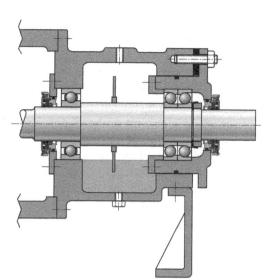

Figure 6.4 A bearing housing accommodating a cartridge on the outboard side. The bearing housing bore is slightly larger than the diameter of the steel flinger disc, making assembly possible. Note that oil pressure/ temperature equalizing passages should have been provided at each bearing!

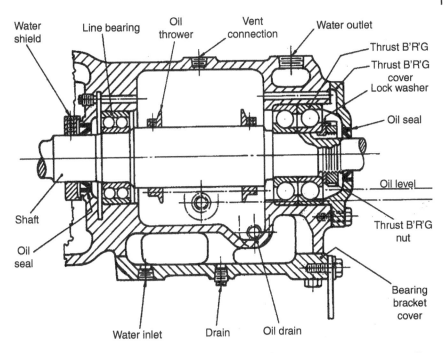

Figure 6.5 This 1960s-vintage bearing housing uses "oil throwers" for keeping the oil at uniform temperature. Note pressure balance holes provided at the top of radial and thrust bearings. *Source:* Worthington Pump Installation and Operating Manual, 1966.

Flexible flinger discs are sometimes used to enable insertion in the simplest bearing housings. In these, the bearing housing bore diameter is smaller than the flinger disc diameter. To accommodate the preferred *solid* steel flinger discs, bearings must be cartridge-mounted (Figures 6.2 and 6.4). With a cartridge design, the effective bearing housing bore (i.e. the cartridge diameter) will be large enough to allow insertion of a steel flinger disc with the needed diameter. Providing such a cartridge will add to the cost of a pump. In the overwhelming majority of cases, the incremental cost of the cartridge design will be low compared to what it costs to repair a pump.

Again, neither Figure 6.2 nor Figure 6.4 show balance holes or passageways that would ensure pressure and temperature equalization. Internal pressure and temperature balance between the central volume of the bearing housing and the spaces between bearings and end caps is essential. This requirement appears to be disregarded in some pump models.

As shown in Figure 6.5 and earlier in Figure 5.5, drilled or similarly machined passageways are needed near either the top or bottom of the bearing housing bore so as to obtain pressure and temperature equalization. In some cases, it would be best to provide equalization passages at both top and bottom. Lack of these passages is one of several explanations for oil leakage and overheated oil. Overheated oil and/or oil contaminated with slivers of O-ring material will result in "black oil."

Note also that the 1960s-vintage bearing housing of Figure 6.5 shows lip seals and water-cooling provisions. Cooling water was deleted from pumps with rolling element bearings in the early 1970s [5]. The lip seals shown here would no longer be acceptable and modern bearing protector seals would be used instead. (See Chapter 8 for details).

Cooling Not Needed on Pumps with Rolling Element Bearings

Cooling is still found in pumps with high speed and heavily loaded rolling element bearings. But cooling of the oil is very rarely needed and often of no benefit in installations with rolling element bearings. Suppose a cooling jacket restricts the bearing outer ring from free thermal growth in all radial directions. But the bearing inner ring heats up and grows, causing bearing internal clearances to vanish. An excessive preload could result. (Chapter 5 dealt with bearing preloads in more detail).

Similarly, immersing cooling coils in the oil will cool not only the oil but also the air in the bearing housing. Such cooling then tends to promote moisture condensation and harmful oil contamination. Therefore, cooling water has been deleted from every pump with rolling element bearings at many Best-in-Class (BiC) locations [1, 5]. Since the late 1970s, there no longer is cooling water on bearing housings of pumps with operating (fluid) temperatures up to and including 740 °F (394 °C) in modern BiC oil refineries.

Because cooling water ports are shown on a pump drawing, the user is led to believe that such cooling is either needed or helpful. Commenting again on Figure 6.5, when it was discovered that cooling is no longer needed, BiC companies began to leave these cooling water drains open. With modern synthetic lubricants and properly selected rolling element bearings, cooling is no longer used in process pumps. So irrespective of lube application method, on *rolling element* bearings cooling will not be needed as long as high-performance synthetic lubricants are utilized.

With sleeve bearings, cooling is used to maintain optimum oil viscosity through close temperature control. This provides a reasonably stable environment for oil rings in pumps equipped with *sleeve bearings*. Circulating systems are the primary choice for large pumps that incorporate these bearings.

Very large process pumps use sleeve bearings and circulating oil systems. In circulating systems, the oil can be passed through a heat exchanger before being returned to the bearing. Pressurization is needed to move oil through filters and exchangers, but the bearing itself is rarely fully pressurized. In some sleeve bearing systems, oil rings are used to lift the oil and deposit it on the shaft surface. In other sleeve bearings, an oil spray from suitably placed nozzles is directed to oil grooves with good effect and very high reliability [1, 6].

Oil Delivery by Constant Level Lubricators

Since the late 1800s, various types of constant level lubricators have been used successfully. Unfortunately, the vulnerabilities of some constant level lubricators are often overlooked. If there are pressure differences between regions in the bearing housings and regions in the lubricator, the two oil levels will differ. Also, whenever caulking is used to secure an oiler bulb or transparent bottle to a supporting component, the caulking will ultimately lose its resiliency or elasticity. Small fissures will develop and rainwater will enter at these fissure locations via capillary action. Therefore, constant level lubricators must be part of conscientious preventive maintenance action. They must be included in a precautionary replacement strategy.

Not all versions of "constant level" lubricators supplied by pump manufacturers will best serve the reliability-focused user. In the widely used pressure *nonbalanced* devices equal or similar to Figure 6.6, the oil level below the reservoir bottle (at the large wing nut that supports the oiler bottle) is contacted by ambient air. But that is so only as long as the pressure in the pump bearing housing to which this device is connected is also at ambient pressure. Should the pressure in the connected pump bearing housing be above ambient, the oil level in the bearing housing would be pushed down and the oil level in the constant level lubricator base (at the tip of the wing nut) would rise.

Look closely at the bearing in Figure 6.6a and note how a small decrease in oil level might stop lubricant from flowing into the bearing. Observe also how the needed mounting location is indicated by the clockwise rotational arrow in Figure 6.6a.

In the *pressure-balanced* device of Figure 6.7 level differences are far less likely to occur. Here, the space surrounding the slanted tube is always at the same pressure as the vapor space in the bearing housing.

Black Oil

Refer back to Figure 6.6a. If this oil level is decreased to a point much below the center of the lowermost rolling element, the risk of oil overheating increases drastically. Many oil formulations typically used in process pumps turn black when

(a)

(b)

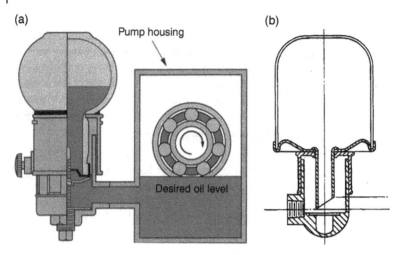

Pump housing

Desired oil level

Figure 6.6 Two different constant level lubricator styles, both are pressure nonbalanced. (a) The wing nut threaded into the vertical rod establishes the oil level. (b) The upper horizontal line represents the oil level. Note that ambient air pressure exists above that line in (b), and above the wing nut in (a). *Source:* Ref. [2].

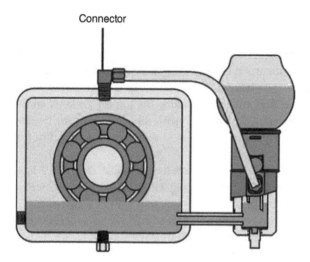

Connector

Figure 6.7 Pressure-balanced constant level lubricator. *Source:* Ref. [2].

overheated. This black oil then shows up in the transparent bowl of constant level lubricators and does so because thermal convection currents cause oil to circulate, and because temperature-dependent cyclic compression and expansion occur in the air space at the top of constant level lubricator bowls. Black oil can show up in the bowl as fine particles of coke.

In situations where the oil reaches the center of the lower-most bearing element, there will be an increase in oil temperature because the rolling elements encounter friction as they move through the lubricant. This temperature increase tends to be excessive at high *DN*-values – another possibility for black oil formation.

Black oil at pump start-up is sometimes the result of an oil ring being wedged into the space between the shaft surface and a long screw that is supposed to limit oil ring travel (shown earlier in Figure 2.1).

It is also possible that inadequate lube oil formulations lack the film strength to maintain a separating oil film between rolling elements and brass or bronze cages at initial pump start-up. Although brass or bronze cages are recommended by API-610 for centrifugal pumps, this particular vulnerability is thought to be greater with copper-containing cage materials than with ferrous metals and suitable plastics.

When there is no longer an adequate oil film, considerable heat will be generated and small portions of the cage material will transfer to the rolling elements. The oil will discolor and contamination can range from minimal to severe.

Additional comments on black oil can be found in Chapter 8, dealing with bearing protector seals. It will again explain that black oil is either overheated oil or lubricant contaminated with slivers of abraded O-ring material.

Lubricant Application As Oil Mist (Oil Fog)

Plantwide oil-mist lubrication systems have proven superior to other methods of lubricating rolling element bearings in process pumps. These systems can supply oil in the form of a dense mist to over a hundred pumps and are limited only by distance traveled [7]. Small oil-mist units are available for serving as lube application modules on up to four pumps (one was shown in Figure 3.3). Used in conjunction with pump and motor bearing housings configured per Figure 6.8, closed oil-mist systems will not emit stray mist fog into the environment.

Oil mist has the appearance of dense fog; it usually consists of a 200000 : 1 (by volume) mixture of dry instrument air and atomized lubricating oil (ISO VG 68 or 100). The mist is produced in a very simple mixing nozzle and then travels through the plant at a pressure of 20–35 in. (500–890 mm) of H2O. The header system is unheated, noninsulated, and at ambient temperature. Individual branch lines come off the top of a header pipe route the mist through a small metering orifice ("reclassifier") to the rolling element bearings of pumps and their electric motor drivers (Figure 6.8).

Because no oil level exists in the oil sump, the pure oil mist method of lubrication is also called dry sump oil mist. With no liquid oil to "lift" from a sump, there no longer are oil rings or flinger discs.

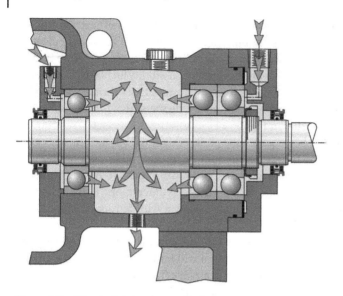

Figure 6.8 Oil-mist lubrication applied to a pump-bearing housing in accordance with long-standing best practices. Note dual mist injection points and use of face-type bearing protector seals to prevent mist from escaping to the atmosphere. *Sources:* Adapted from Bloch and Shammim [7]; Bloch [7, 8].

Electric motor drivers and entire nonrunning (standby) pump sets are included in plantwide oil-mist systems. Dry sump (pure) oil mist applied per Figure 6.8 clearly eliminates every one of the previously discussed problems with oil rings. It also avoids or eliminates and supersedes issues with defective constant level lubricators and missing balance holes in bearing housings [7–9].

Cost justification calculations show payback periods typically in the range from six months to three years for many oil mist installations. The more rapid payback is associated with, among other things, fewer bearing failures resulting in a reduced number of fire incidents in process plants [1]. The less-rapid payback is for environmentally friendly "closed" oil-mist systems.

All things considered and for process plant applications, dry sump (pure) oil mist is by far the most reliable, predictable, and least-risk means of providing trouble-free bearing lubrication for process pumps. There are no longer any oil rings, and black oil is not experienced with oil mist applied per Figure 6.8. Note that face-type bearing protector seals are used in up-to-date oil-mist lubricated bearing housings. The through-flow routing in Figure 6.8 complies with best-available practices. These practices were first implemented by Shell and Exxon Chemical plants in the early 1970s and have been successful ever since then.

The reliability and availability of modern oil-mist systems exceed other methods of pump lubrication. At one plant and over a period of 14 years, a single qualified contract worker serviced the 17 systems by visiting the facility one day each month. In this 14-year time period, there was only one malfunction; it involved a defective float switch in the bulk oil container of one of the 17 oil-mist generator consoles. The incident caused a string of pumps to run without a functioning mist supply system for eight hours. Still, there were no bearing failures. In this plant, the combined long-term availability and reliability of the oil-mist systems were calculated to have reached 99.99962% [7, 8].

The advantages and disadvantages of oil mist lubrication as compared to traditional liquid oil sump (oil bath) lubrication are summarized as follows:

Advantages
- Represents automated lube application for pump, driver, and standby set.
- Reduces bearing failures by 80–90%.
- Lowers bearing operating temperatures by typically 10–20 °F (~6–12 °C).
- Continuously removes bearing wear particles.
- Slight positive system pressure eliminates contaminant entry.
- Reduces energy costs by 3–5%.
- Reduces oil consumption by about 40%.
- Greatly reduced maintenance intensity because oil mist generator and its associated equipment contain no moving parts

Disadvantages
- Capital investment
- Cost of compressed air

Desiccant Breathers and Expansion Chambers

Oftentimes desiccant-breather combinations seem to address the *symptom* of a housing-internal pressure-related shortcoming and might even represent a fix, in some isolated instances. No such breathers are needed in closed bearing housings or if pressures are equal in front and behind bearings.

If used, desiccant-breathers are maintenance items that must be considered in the budget. These elements or containers are filled with a chemical that absorbs moisture. The container is often made of a transparent polycarbonate or similar plastic material.

Some users also believe in expansion chambers (Figure 6.9) and have occasionally modified the tops of pump-bearing housings so as to allow installation of both a desiccant-breather and an expansion chamber. Expansion chambers serve no technically advantageous purpose in vented pump-bearing housings.

Figure 6.9 Expansion chamber intended for process pump-bearing housings. *Source:* www.acklandsgrainger. com [2].

In fully sealed bearing housings, the pressure increase is a function of temperature rise. The degree to which bearing housing pressures are reduced by expansion chambers is then a function of the volumetric ratio of the chamber and the bearing housing volume. (Recall the combined gas laws: $P1V1/T1 = P2V2/T2$).

If a pump has a relatively large bearing housing volume and/or low temperature rise, adding a small expansion chamber will serve little purpose. Applying the gas law equation will prove it.

What We Have Learned

There are truly hundreds of combinations of oil application, bearing lubricant bypass possibilities, pressure equalization options, and venting arrangements. Examining a bearing housing cross-sectional drawing is the minimum requirement for a competent failure analyst. Physically examining a bearing housing and its bearings is even better. (Remember: you get what you inspect, not what you expect).

Here is a recap relating to shaft diameter and shaft rpm interaction:

- On sump-lubricated pumps with DN [dn] values (shaft diameter in inches [mm] times rpm) up to 6000 [$dn \sim 160\,000$], allow lube oil to reach the center of the rolling elements at the 6 o'clock position.
- Total flooding of pump bearings (regardless of DN or dn value) may result in excessive heat generation.
- For pumps with DN [dn] values over 6000 [$dn \sim 160\,000$], the oil level may have to be lowered and oil rings or flinger discs may have to be chosen by the user or pump manufacturer.
- Many oil rings are unstable and skip at DN [dn] values over 6000 [$dn \sim 160\,000$]; some Best-of-Class companies will not use them above DN [dn] values over 8000 [$dn \sim 213\,000$].

Among the elusive reasons why pump bearings fail, we will find these four:

- an occasional unreliable *wet sump* oil-mist application
- dry sump oil mist introduced in a nonoptimized manner (i.e. arrangements that do not conform to the 10th Edition of API-610, see [10])

- unexpected pressure drops through desiccant containers
- misunderstandings as to what bearing housing expansion chamber will do and will not do

We learned to pay attention to avoiding pressure gradients in the vicinity of bearings. Certain angular bearing cages act as impellers and create windage effects that make pressure balance within the bearing housing difficult.

Next to the vastly superior dry sump oil mist, installing a bull's-eye-style-sight glass and omitting the constant level lubricator is the closest acceptable alternative. A properly dimensioned flinger disc and modern bearing protector seals (Chapter 8) would be part of this closest acceptable alternative.

An inexpensive "all purpose" or "standardized" lubricant will give mediocre results at best. In the end, buying the right lubricant formulation – even at a premium price – will be the most cost-effective option for reliably lubricating process pumps [11]. Involving a lubricant provider staffed by application engineers [8] will be advantageous. Application engineers are informed readers who know best practices.

Many different variables influence oil levels in pump-bearing housings, windage, and venting are among them. No statistics are available that allow definitive cataloging or linking of the many factors. But doing all things right is the path to satisfactory pump operation. While a small deviation from the norm can sometimes be tolerated, allowing several parameters to reach their individual limits of acceptability will always have a negative effect on pump reliability.

References

1 Bloch, Heinz P., and Budris, Alan; "*Pump User's Handbook: Life Extension*", 2nd Edition, Fairmont Publishing Company, Lilburn, Georgia, 2006 (ISBN 0-88173-627-9).
2 TRICO Manufacturing Corporation, Pewaukee, WI; Commercial Literature and www.tricocorp.com.
3 SKF USA, Inc.; "Bearings in Centrifugal Pumps", 2nd Edition, Publication 100-955, 1995.
4 Bradshaw, Simon; "Investigations into The Contamination of Lubricating Oils in Rolling Element Pump Bearing Assemblies", Proceedings of the 17th International Pump User's Symposium, Texas A&M University, Houston, Texas, 2000.
5 Bloch, Heinz P.; "*Machinery Reliability Improvement*", 3rd Edition, Gulf Publishing Company, Houston, Texas, 1998 (ISBN 0-88415-661-3).
6 Bloch, Heinz P.; "Inductive Pumps Solve Difficult Lubrication Problems", *Hydrocarbon Processing*, September 2001.
7 Bloch, Heinz P., and Shamim, Abdus; "*Oil Mist Lubrication – Practical Application*", Fairmont Press, Lilburn, GA, 30047, 1998 (ISBN 088173-256-7).

8 Bloch, Heinz P. (2021) *Optimized Equipment Lubrication: Conventional Lubrication, Oil Mist Technology, and full Standby Protection*, DeGruyter Publishing, Berlin/Germany, ISBN 978-3-11-074934-2.

9 Bloch, Heinz P., and Geitner, Fred; *"Machinery Component Maintenance and Repair"*, 3rd Edition, Gulf Publishing Company, Houston, Texas, 2005 (ISBN 0-7506-7726-0).

10 API-610; *"Centrifugal Pumps"*, 10th Edition, American Petroleum Institute, Alexandria, VA, 2009.

11 Bloch, Heinz P.; *"Practical Lubrication for Industrial Facilities"*, 2nd Edition, Fairmont Press, Lilburn, GA, 30047, 2009 (ISBN 088173-579-5).

7

Lubricant Types and Key Properties

Lubricant Viscosities

Viscosity is by far the most important property of the lubricants applied to process pump bearings and practical texts deal with the issue in great detail [1]. In general, thicker oil films will give better pump bearing protection than thinner oil films. For process pumps with rolling element bearings, ISO Grade 68 lube oils will allow higher operating loads. They are generally preferred over ISO Grade 32 lubricants, but note that oil rings are designed for ISO VG oils.

Not all application methods are possible with ISO viscosity grades higher than VG 32. How to capture the benefits of thicker lubricants without actually using them or how to best apply ISO VG 68 and thicker oils will be the subject of discussion later in the chapter.

Figures 7.1 and 7.2 allow quick determination if a particular lubricant selection is in the right range. Suppose we were dealing with a 1800 rpm pump and its bearings had bore diameters of 65 mm (dimension d) with bearing outer diameters (D) of 120 mm.

Since $dm = 0.5 (d+D)$, in this case $dm = 92$ mm. Enter the horizontal x-axis of Figure 7.1 at this calculated mean dimension and plot upward to intersect an imaginary 1800 rpm diagonal line – just above the 2000 rpm diagonal. From there, move to the left and read off ~11 mm^2/s (a measure of viscosity that is more commonly called 11 centistokes, generally abbreviated as cSt).

We have now established that, in this example, the minimum kinematic viscosity required to give adequate protection at operating temperature is 11 cSt. Since oils become thinner when heated up and if our operating temperature is very high, we realize we should have selected thicker oil. As this thicker oil then

Pump Wisdom: Essential Centrifugal Pump Knowledge for Operators and Specialists, Second Edition. Robert X. Perez and Heinz P. Bloch.
© 2022 The American Institute of Chemical Engineers, Inc. Published 2022 by John Wiley & Sons, Inc.

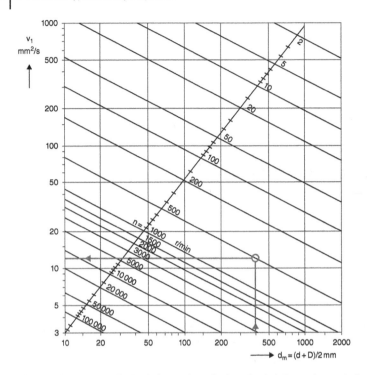

Figure 7.1 The required minimum (rated) viscosity (v_1) depends on shaft rpm and bearing size. *Source:* SKF USA, Kulpsville, Pennsylvania [2].

reaches a higher operating temperature, it will, hopefully, not "thin out" to a viscosity below 11 cSt.

Continuing on Figure 7.2, we must now either assume a certain bearing operating temperature, say, 70 °C or 158 °F. Entering the horizontal x-axis scale at 70 °C and moving upwards to ISO VG 68 would allow us to read off 20 cSt on the vertical scale – close to twice what we need. (Important: We should always ascertain that our oil delivery system works with this thicker-than-needed oil). We might pick ISO VG 32 and, as long as our actual bearing operating temperature does not exceed our assumed temperature of 70 °C, we would now read off (from the vertical y-scale) an operating viscosity of 11 cSt – just right.

We could also use Figure 7.2 by entering the vertical y-scale at the required 11 cSt and, after intersecting a particular viscosity grade oil, plot down to read off the maximum allowable bearing operating temperature on the horizontal x-scale. Thus, intersecting at, say, VG 68, would tell us that this oil could be allowed to reach 92 °C and still satisfy our viscosity requirement of 11 cSt.

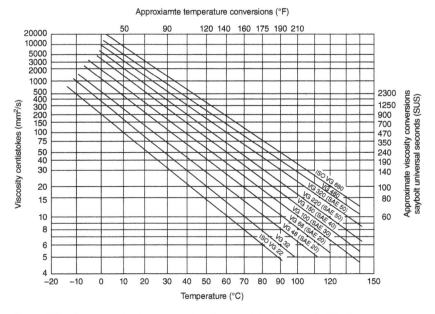

Figure 7.2 For a required viscosity (vertical scale), the permissible bearing operating temperatures (horizontal scale) increases as thicker oils are chosen (diagonal lines).

Thicker film oils are easily and reliably applied with the oil mist method briefly summarized in Chapter 6. However, thick oils can be quite difficult to apply with only the oil rings shown earlier in Figure 5.3.

Running a pump one might ultimately achieve an operating temperature that allows a certain thick oil to flow nicely. But what if, at startup, the initial operating temperature is quite low and the oil will not flow? That's what often happens when someone buys a standard *multipurpose oil* without really thinking things through.

Recall or refer back to Figures 5.5 and 5.18 which showed lube applications with the oil level reaching to the center of bearing elements at the 6-o'clock position. In these instances, ISO Viscosity Grade 68 allows operation in a relatively wide range of ambient temperatures.

However, in applications using oil rings and with oil levels reduced below the periphery of the lowermost bearing component (see Figures 5.3 and 5.19), ISO Grade 32 lubricants may have to be used simply because not all oil rings will work with thicker oil viscosities. At the DN values typically encountered in process pumps, most oil rings function better in ISO Grade 32 lubricants than they would in ISO Grade 68 oils.

Here are a few general guidelines worth considering:

- Using a mineral oil would generally require oil to be changed every 6–12 months. With a clean, premium grade synthetic lubricant one would typically extend oil change intervals to about 24 months.
- ISO Grade 32 *mineral* oils are often considered too thin for pump bearings. They rarely qualify for long-term, risk-free use in pumps equipped with rolling element bearings in typical North American and European ambient conditions. But simply switching to ISO Grade 68 *mineral oils* will be risky for bearings that are fed by oil rings.
- Appropriately formulated with the right base stock and with proprietary additives, ISO VG 32 *synthetics* are quite acceptable from film strength and film thickness points of view [1, 3]. In fact, the performance of some ISO VG 32 synthetics duplicates that of ISO VG 68 *mineral* oils. These superior ISO VG 32 synthetics excel by simultaneously satisfying the requirements of sleeve bearings and rolling element bearings [4].
- Superior synthetics achieve high film strength through proprietary additives. So there can be significant differences in the performance of two lubricants of the same viscosity and using the same base stocks. Only one might be suitable for highest reliability services.
- The notion that one oil type or viscosity suits all applications is rarely correct and is easy to disprove. Similarly, no fixed or particular oil ring geometry is ideally suited for all oil types and viscosities. Custom-designed oil rings may be required to work with the thicker oils at certain high shaft DN values.

When and Why High Film Strength Synthetic Lubricants Are Used

Quality lubrication includes sound and risk-free application method, proper lube quantity, appropriate oil type and viscosity, properly storing and handling the lubricant, attending to bearing housing contamination issues, and implementing appropriate oil change intervals.

To summarize good lubrication practices: we must choose the right oil, take proper care of it, and change it before bearings are harmed. Improvements in lubricant quality can only be achieved by utilizing oils with superior lubricating properties. These would be premium synthetics.

Even among prominent synthetic lubricants, oil performance can vary greatly based on the amount and composition of additives in the oil. For process pump bearing lubrication, at least one company combines synthetic base oils including poly-alpha-olefin (PAO) and dibasic ester base stocks with advanced additive chemistry so as to realize greater film strength [4].

Numerous incidents have been documented where advanced lubrication technology has significantly improved pump reliability. In the majority of cases, advanced lube technology with its often more favorable (lower) coefficient of friction results in reduced bearing operating temperatures. Microcracks in bearing surfaces cause increased noise and vibration; suitable high film strength oils will fill these microcracks. This then lowers noise intensity and reduces vibration severity [1, 3].

High film-strength lubricants also lessen the probability of lube oil darkening during the running-in period of bearings with brass or bronze cages. There have been reported instances of high frictional contact during the initial run-in period of the copper-containing material recommended – for its other qualities – by API-610. If the net axial thrust action on one of the two back-to-back oriented bearings causes it to become unloaded, it may skid (see Figure 5.8).

The risk of lube oils darkening during the run-in period of such pumps is reduced through the use of high film-strength synthetic lubes. To be fair, this risk could also be reduced by insisting on impeccable installation techniques and the selection of bearings with cages made of advanced high-performance polymers (see Chapter 5 regarding temperature limits for plastic cages).

Whatever the differential cost of a quart (or liter) of high film-strength synthetic, it is insignificant compared to the value of an avoided failure incident on critical, nonspared refinery pumps.

Critically important pumps, pumps in high temperature service, and pumps that have failed more often than others in the plant's pump population should, therefore, be lubricated with high film-strength synthetic oils.

On pumps where a problem is in progress, changing to a superior synthetic is highly recommended. If access to the sump drain is safe while the pump is in service, the present oil can be drained while such pumps are on-line and running. Many superior synthetics are compatible with the oil presently used in a particular pump.

Switching to superior film strength synthetic lubricants would give immediate payback. Virtually every cost justification calculation indicates unusually large benefits for employing these lubes on problem pumps.

Of course, there are certain pump bearing or lube degradation problems that have nothing to do with the lubricant type. In those instances, nothing will be gained by changing over to better oils. (There is never a substitute for responsible and accurate failure analysis, see Chapter 16).

Lubricants for Oil Mist Systems

Pure oil mist lubrication (Chapter 6) eliminates the need for either oil rings or flinger discs. No liquid oil sumps are maintained in the bearing housings; hence, the term "dry sump" is often used to describe modern oil mist lubrication.

ISO VG 68 and VG 100 mineral or synthetic oils are used, although properly formulated ISO VG 32 *synthetics* (but not mineral oils) will serve the majority of pump bearings and also virtually all types of rolling element bearings in electric motors. Decades of experience on thousands of pumps and electric motors attest to the viability and cost-effectiveness of modern plant-wide oil mist systems. Typical payback periods when using oil mist on problem pumps have generally been less than one year.

As shown in the bar chart of Figure 7.3, bearing friction can be reduced by switching to different oils, or going with a different lube application method, or by switching both lube application method and oil type. Five different modifications were closely examined in a cooperative effort involving a multinational lube oil producer and prominent bearing manufacturer. The results are plotted and percentage reductions in bearing friction displayed on the vertical scale of Figures 7.3 and 7.4.

What We Have Learned

- If repeat failures occur in process pumps, there are overwhelming odds of several small deviations combining. Once several (in themselves tolerable) deviations combine, just one more blip becomes the proverbial "straw that breaks the camel's back." In other words, once several deviations exist, one additional deviation will often cause a serious failure.

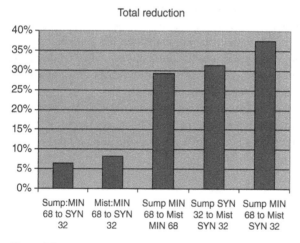

Figure 7.3 How changes in lube application, oil types, and lube viscosities tend to affect percentage reductions in bearing friction; these are displayed on the vertical scale. MIN = mineral oil; SYN = synthetic oil. *Source:* Royal Purple Ltd., Porter, Texas [5].

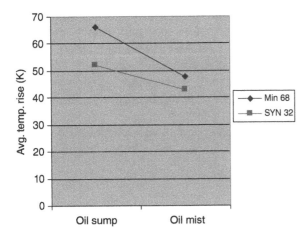

Figure 7.4 Illustrates an important difference between mineral oils and synthetics, The two lubricants shown here have similar performance characteristics. However, the Synthetic ISO VG 32 lubricant excels in removing heat from bearings. *Source:* Refs. [3, 6, 7].

- Using a "normal" lubricant is fine as long there is no excessive axial load on the bearings. With a very high axial load, superior quality oils will be mandatory.
- Having an oil ring with an eccentricity slightly greater than the normally allowed 0.002 in. (0.05 mm) *seems* quite acceptable, but only until perhaps the out-of-horizontality of the shaft system exceeds a certain value, or until there is – additionally – a certain deviation from a normally close range of lube oil viscosity, or until the increases in generally tolerable cavitation, or bearing-related vibration or whatever else, converge to cause calamity.

Make it your business to understand these facts and act on this understanding. You do have it in your power to avoid process pump failures.

One has to be consistently inside the acceptable ranges of dimensional, material composition, fabrication-specific, application-related, and a host of other parameters and disciplines and ingredients. Adherence to sound specifications is not difficult once a proper mindset is cultivated. The difficulty is in cultivating the mindset.

References

1 Bloch, Heinz P.; "*Practical Lubrication for Industrial Facilities*", 2nd Edition, Fairmont Press, Lilburn, GA 30047, 2009 (ISBN 0-88173-579-5).
2 SKF USA, Kulpsville, Pennsylvania; "General Catalog", (Engineering Section), 2008.

3 Bloch, Heinz P., and Allan, Budris; *"Pump User's Handbook: Life Extension"*, 3rd Edition, Fairmont Press, Inc., Lilburn, GA, 30047, 2009 (ISBN 0-88173-627-9).

4 Morrison, F.R., Zielinsky, J., and James, R.; "Effects of Synthetic Fluids on Ball Bearing Performance", ASME Publication 80-Pet-3, February, 1980.

5 Royal Purple Ltd., Porter, Texas (2004) "Marketing Literature", 2000–2004.

6 Wilcock, Donald F., and Booser, Richard E.; *"Bearing Design and Application"*, McGraw-Hill, New York, NY, 1957.

7 Villavicencio, E.; "Basic Elements of Tribo-Thermo-Dynamics", presented at the First International Symposium on Reliability and Energy Tribo-Efficiency, Mexico City, June 2002.

8

Bearing Housing Protection and Cost Justification

Both oil and air fill the bearing housing. As either gets warm and expands, or cools and contracts, an interchange of the internal air takes place with the surrounding or ambient air. The interchange is called "breathing."

To stop this breathing and resulting contamination, the breather vents on the housings shown earlier in Figure 2.1 should be plugged and suitable bearing protector seals[1] installed where the shaft protrudes through the housing.

Inexpensive lip seals are sometimes used for sealing at the bearing housing, but lip seals typically last only about 2000 operating hours – three months [1]. When lip seals are too tight, they tend to cause shaft wear similar to that shown in Figure 8.1 and, in some cases, lubricant discoloration ("black oil," [2]) and contamination. Once lip seals have worn and no longer seal tightly, oil is lost through leakage. The API-610 standard for process pumps disallows lip seals and calls for either rotating labyrinth-style or contacting face seals.

The bearing housing protector seal in the lower right portion of Figure 8.1 incorporates a small and a large diameter dynamic O-ring. This bearing protector seal is highly stable and not likely to wobble on the shaft; it is also field-repairable. With sufficient shaft rotational speed, one of the rotating ("dynamic") O-rings is flung outward and away from the larger O-ring. The larger cross-section O-ring is then free to move axially and a microgap opens [2].

When the pump is stopped, the outer of the two dynamic O-rings will move back to its standstill position. At standstill, the outer O-ring contracts and touches the larger cross-section O-ring. In the purposeful design of Figure 8.1, the larger cross-section O-ring touches a relatively large contoured area. Because Contact Pressure = Force/Area, a good designs aims for low pressure. A good design,

---

1 The terms "bearing housing protector seal," "bearing protector seal," and "bearing isolator" are used interchangeably.

Pump Wisdom: Essential Centrifugal Pump Knowledge for Operators and Specialists, Second Edition. Robert X. Perez and Heinz P. Bloch.
© 2022 The American Institute of Chemical Engineers, Inc. Published 2022 by John Wiley & Sons, Inc.

(a)   (b)

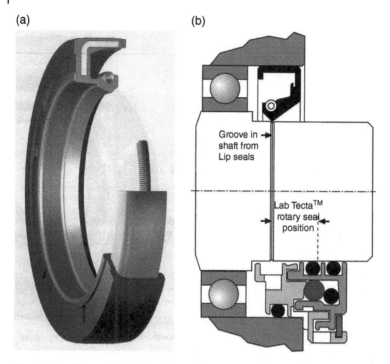

Groove in shaft from Lip seals

Lab Tecta™ rotary seal position

Figure 8.1 Lip seals (a and top b) tend to wear and have typically only a three-month operating life; replacement with a high-reliability rotating labyrinth seal (below, b) is possible without first having to repair damaged shaft surface.

Figure 8.1, will differ greatly from the model shown in Figure 8.2a, where contact with the sharp edges of an O-ring groove will cause O-ring damage.

Rotating bearing housing protector seals ("bearing protectors") are classed into the main categories *contacting* and *noncontacting*. Contacting styles include (i) plain lip seals, (ii) lip seals engaging a tapered groove labyrinth, and (iii) the various face-contacting types. Low-cost noncontacting original equipment manufacturer (OEM) bearing protectors are often of the simplest labyrinth configuration (Figure 8.2a,b). A least-risk configuration is shown in Figure 8.2c.

Noncontacting Bearing Protector Seals

To recap: By definition, *noncontacting* implies at-speed operation with a very small gap between rotating and stationary sealing components. Optimum gaps are found in advanced rotating labyrinth seals, Figure 8.2c.

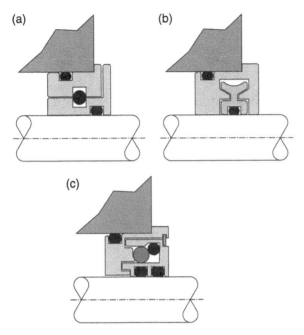

Figure 8.2 Noncontacting bearing protector seals: (a) plain labyrinth – its O-ring risks being damaged by sharp grooves; (b) sealing lip engaging a tapered groove, a high drag effect is possible; (c) dual dynamic O-ring and dual clamping rings as found in a highly reliable, field-repairable design.

In noncontacting bearing protector seals, the *rotating* component typically has a complex outer profile. This profile is located adjacent and in close radial and axial proximity to the complex inner profile of a *stationary* component.

Together and in theory, these complex profiles form a tortuous path that prevents inward or outward passage of unwanted materials or fluids. Protector seals are more likely to fail if the design places a dynamic O-ring close to razor-sharp circumferential edges or grooves (Figure 8.2a).

Note that bearing protector seals often differ in basic design and detailed geometry. The details may be quite important, see Figure 8.3. Here, the reliability professional must consider the effects of overhung mass and a single O-ring used for clamping. Rotor tilt will cause the squeezed clamping ring to become "kneaded" and the rotor to walk.

Figure 8.3 thus points to an elusive root cause of failure. In this instance, O-ring debris can end up in the bearing housing, causing the well-documented appearance of "black oil." (Ref. [3] and several other chapters in this book mention "black oil" – please see index).

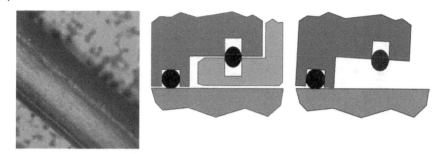

Figure 8.3 Inadequate shaft clamping, off-center mass, and O-ring contact with sharp edges can cause degradation and unforeseen contamination of lube oil. *Source:* Bloch [3].

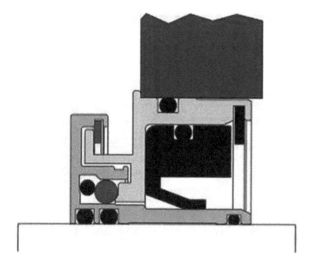

Figure 8.4 Advanced hybrid bearing protector seal.

Contacting Bearing Protector Seals

In the second category, *contacting* bearing protectors, users have also employed technically advanced hybrid seals (Figure 8.4) and face-contacting magnetic seals (Figure 8.5). In magnetic seals, the closing force is supplied by a number of floating permanent rare-earth magnets that exert a pulling force on steel Z-rings. Mounted in these Z-rings are carbon faces which then contact an O-ring mounted single piece rotor. Face-contacting seals are occasionally used in pump bearing housings that receive lubrication from closed-loop oil mist systems [4].

Both configuration-related categories of bearing protectors are available as designs that can accommodate angular shaft-to-housing misalignment (Figure 8.6). Variants for steam turbines have also proven highly successful.

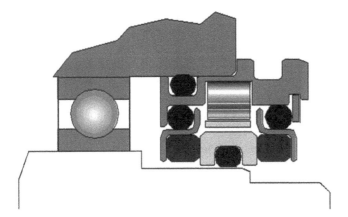

Figure 8.5   Dual-face magnetic bearing protector seal.

Figure 8.6   Self-aligning and extended axial
movement bearing protector seal.

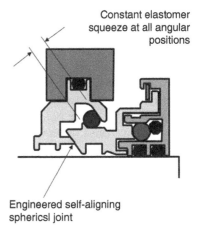

Constant elastomer
squeeze at all angular
positions

Engineered self-aligning
sphericsl joint

How Venting and Housing Pressurization Affect Bearing Protector Seals

If in nonvented bearing housings, the pressure starts to increase with rising temperature, lip seals will release the excess pressure by lifting off. To release pressure, seal orientation is important. Lip seal frictional drag (and wear) would increase as housing pressure rises in Figures 8.1 and 8.4. In contrast, the microgap in the dual dynamic O-ring protector seals in Figures 8.1, 8.2c, 8.4, and 8.6 would act as pressure relief passages.

Pressurizing a bearing cavity equipped with a single or dual-face magnetic seal (Figure 8.5) will add to the closing force of the rod magnets facing the left side of the rotor. The left Z-ring is close to the bearing, the Z-ring to the right of the rotor is close to the ambient environment. Housing-internal pressures up to 2 psig (13.8 kPa) over ambient are allowed in either single or dual-face magnetic seals, with the seal manufacturer's concurrence.

Pressure increase in nonvented bearing housings can be calculated using the combined gas law equation. This law says the ratio between the (absolute) pressure-volume product and the (absolute) temperature of a system remains constant. The pressure rise resulting from a known or assumed temperature rise in the same volume is simply $P_2 = P_1 T_2 / T_1$.

Since operating temperatures are rarely more than 100 °F (~56 °C) above standstill temperatures, the resulting pressure rise would stay below 2 psi. This Δp is too small to justify installing expansion chambers on most process pumps. (See Chapter 6 for comments on desiccant breathers and expansion chambers).

Cost Justification Overview

Lip seals will seal only while the elastomer material (the lip) makes full sliding contact with the shaft. Operating at typical shaft speeds on process pumps, lip seals show leakage after about 2000 operating hours. To prevent contaminant intrusion, one would have to replace lip seals before they fail – twice a year for a certainty. In sharp contrast, modern rotating labyrinth seals incorporating the features seen in the lower right portion of Figure 8.1 would have a life typically exceeding four years [5]. Let us look at two cost comparisons:

- *Scenario 1 – Replace lip seals before they fail*: To avoid shaft fretting, moisture intrusion, and premature bearing failure (assuming labor and materials to remedy a bearing failure costs US$ 6000), one would have to replace a US$ 20 lip seal before it fails, say, twice a year. Labor (US$ 500 per event) and materials (2 per pump) = US$ 1080/yr.

  Alternatively, assume (just for the case of illustration) we replace two US$ 200 modern dynamic O-ring rotating labyrinth seals after four years of operation. Labor: US$ 125/yr, material: US$ 100/yr; Total = US$ 225/yr; Benefit to cost ratio: 1080/225 = 4.8.

- *Scenario 2 – Run lip seals to failure*: Lip seals are allowed to degrade and bearings fail after two years. Assume no production outage time, but the repair costs the plant US$ 6000 = US$ 3000/yr.

  Again and as an alternative, assume (for the case of simple, ultraconservative illustration) we replace two US$ 200 modern dynamic O-ring rotating labyrinth seals after four years of operation. Labor: US$ 125/yr, material: US$ 100/yr; Total = US$ 225/yr; Benefit to cost ratio: 3000/225 = 13.3.

Total cost of ownership and not just the as-purchased cost of the seal is very important. More detailed cost justifications always include maintenance expenditures for oil changes. Note that labor, material, waste oil disposal, and other costs would have to be included. This would shift the picture even more in favor of modern bearing protector seals.

Advanced Bearing Housing (Bearing Protector Seal) Summary

Lip seals have their place in disposable appliances and in machines which, for unspecified reasons, must frequently be dismantled. Lip seals do not measure up to the expectation of most intermediate and heavy-duty duty pump users.

Advanced rotating labyrinth seals are much preferred for reliable long-term operation of process pumps. In the span of several years between their market introduction and release of this book, *none* of the well-engineered models shown in Figure 8.1 had failed in operation and tens of thousands were in operation at release time. Many were narrow-face types with single clamping O-rings.

Long life was demonstrated for both narrow and wide versions of this product. There is little (if any) risk of rotor-stator contact because the mass-symmetrical rotor is clamped to the shaft with two O-rings. The two-ring shaft clamping method imparts excellent stability and resists "rotor wobble."

Of course, bearing protector seals serve no purpose

- if oil contamination originates with oil ring inadequacies or
- if used with unbalanced oilers (Chapter 6) or
- if the oil is not kept at the proper oil level or
- if the design causes or increases the risk of "black oil" formation [2, 3].

On the other hand, soundly engineered bearing protector seals excel by incorporating three attributes:

1) The assembly uses two O-rings (and not just one) to clamp the rotor to the shaft;
2) The rotor mass center is located centrally between these two O-rings and the rotor mass is *not* overhanging asymmetrically or outside of the two O-rings.
3) To qualify as soundly engineered bearing protector products and assuming they incorporate dynamic O-rings as sealing elements, these dynamic O-rings should act on a generously dimensioned, contoured surface and *not* on a sharp edge or groove.

What We Have Learned

Bearing housing protector seals can greatly improve both operating life and reliability of process pumps by safeguarding the cleanliness of the lubricating oil. As a rule, advanced products are estimated to extend pump bearing life fourfold.

Yet, not all bearing housing protector seals (bearing isolators) perform equally well. The user-purchaser must become familiar with the working principles of different offers and steer clear of vulnerable old-style configurations.

Thoroughly well engineered products will not allow O-rings to contact sharp-edged components. They are superbly designed, with no chance of O-rings contacting sharp-edged grooves or other components. Although some of these designs are field-repairable and could be rebuilt by simply replacing a set of O-rings, thousands of these bearing housing protector seals have been in flawless service for many years. Few, if any, ever needed O-ring replacement.

References

1 Brink, R. V., Czernik, D. E., and Horve, L. A.; "*Handbook of Fluid Sealing*", McGraw-Hill, New York, 1993 (ISBN 10: 0070078270).

2 Bloch, Heinz P., and Budris, A.; "*Pump User's Handbook: Life Extension*", 3rd Edition, Fairmont Publishing Company, Lilburn, Georgia, 2006 (ISBN 0-88173-629-9).

3 Bloch, Heinz P.; "Fighting Black Oil: How Alternatives to Oil Rings May Help Prevent Degradation", *Turbomachinery International*, Jan/Feb 2009 (Part 1), also March/April 2009 (Part 2), 2009.

4 Bloch, Heinz P., and Shamim, Abdus; "*Oil Mist Lubrication–Practical Application*", Fairmont Press, Lilburn, GA, 30047, 1998 (ISBN 088173-256-7).

5 Bloch, Heinz P., and Rehmann, Chris; "Well Worth the Cost–Bearing Housing Seals and Value of Reducing Oil Contamination", *Uptime Magazine*, Aug/Sept 2008.

9

Mechanical Sealing Options for Long Life

Mechanical seals are components that keep the pumped fluid (or "pumpage") in the pump casing. These seals are installed at the location where the shaft enters or leaves the casing and their purpose is to prevent leakage. There are many hundreds of styles and sizes and configurations of seals. All use the underlying principles of stationary and rotating face combinations (Figure 9.1). Some are simple and inexpensive; others are complex and deserve special attention.

Conventional mechanical seals often apply spring pressure or some other closing force to the face of the rotary unit in Figure 9.1. However, many mechanical seals are designed with a single coil spring, several small coil springs, or pleated bellows applying a closing force to the stationary seal part. They are then called *stationary* mechanical seals. Application of spring-loaded closing force to the stationary seal part is advantageous when there is known shaft deflection and in shaft systems that operate at high peripheral velocity. Also, designs that place the springs away from the process fluid are generally preferred over designs that allow process fluid to contact the springs. All seal assemblies need some kind of vent provision, be it a port "A" or a simple vent passage "B" (Figure 9.1).

Still Using Packing?

Since about 1950, mechanical seals have almost totally replaced the more maintenance-intensive compression packing previously used in process pumps (Figure 9.2). But suppose you are presently in the process of converting to

Pump Wisdom: Essential Centrifugal Pump Knowledge for Operators and Specialists,
Second Edition. Robert X. Perez and Heinz P. Bloch.
© 2022 The American Institute of Chemical Engineers, Inc. Published 2022 by John Wiley & Sons, Inc.

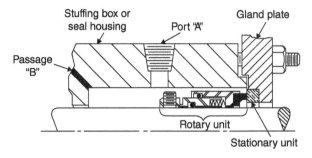

Figure 9.1 Operating principle and essential nomenclature of mechanical seals.

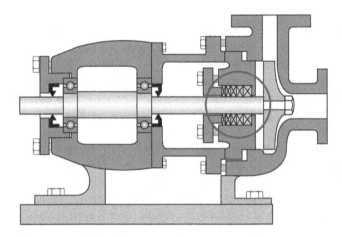

Figure 9.2 Compression packing used in a simple centrifugal pump.

mechanical seals. In that case and until conversion work is completed, use this installation and servicing sequence:

Distinguish between different packing styles and materials, be sure to use the right one for the application

Preferably use a mandrel to properly wrap and cut new packing

Use the right tool to remove old packing

Remove corrosion products from shaft or sleeve surface

Coat ring surfaces with an approved lubricant

Insert one new ring at a time; then seat each ring using a proper tool

Install packing gland, initially only finger-tight

Operate for run-in, initially allowing an uninterrupted stream of sealing water leakage

Tighten sequentially to allow flow of 40–60 drops per minute (inadequate leakage flow generates much heat and burns up packing)

General Overview of Mechanical Seals

Working with a competent seal manufacturer avoids mistakes and adds value. Reliability-focused users never view mechanical seals as a cheap standard commodity. These users select only experienced seal manufacturers and cultivate relationships based on mutual trust. Long-term customer service and consistent application of thorough engineering skills benefit both parties and are of much greater importance than short-term returns obtained from initially paying a low price. A single failure incident often causes price-related gains to vanish in a flash – both literally and figuratively.

Seal flush piping arrangements are often needed to create the most appropriate seal environment. The ones shown in this chapter and many other variants generally belong to two main groups:

1) "Flush" – Clean, cool liquid is injected into the seal chamber ("F1", Figure 9.16) to improve the operating environment. A second, smaller port is provided and normally oriented downward. It often serves as a leakage observation or vent opening (ports "Q/D", Figure 9.16).
2) "Barrier or Buffer" used with dual seals – A secondary fluid is fed to the space between two sets of mechanical seals to prevent ambient air from contacting the pumped fluid, to improve seal cooling, or to enhance safety.

It is worth mentioning that not all mechanical seals require special external flush arrangements. Still, for long and trouble-free life the mating faces shown in Figure 9.1 must somehow be separated or cooled. The very small gap between faces must be taken up by either liquid or gas. Different flush plans (flush schematics) accomplish that; they vary because they must accommodate different fluid parameters, conditions, and properties. All flush plans are described in vendor literature and industry specifications such as API-682 and ISO 21049. A few of them are shown in Figures 9.3–9.8; the ones we have selected here use American Petroleum Institute (API) plan numbers and are representative of the many different piping or flush plans available [1].

Note that the flush piping in most plans enters or exits at the top of the seal chamber which then also allows vapors to be vented. In the few instances, where no such piped venting possibility exists, the pump user may have to drill a 0.125 in. (3 mm) diameter passageway (labeled "B" in Figure 9.1) at an angle of 15°–45° from the top corner of the seal chamber into the fluid space directly behind the impeller.

In preexisting pumps, certain upgrades or conversion to another (more modern) flush plan configuration are often advantageous. Where energy conservation issues and operating cost savings are deemed important the newer seal configurations must be considered [2].

Recirculation of a product side stream (Plan 11, Figure 9.3) is common. The side stream should have a pressure of approximately 25 psi (173 kPa) higher than the pressure directly behind the pump impeller [3].

Axial holes have been drilled in the impeller discs of Figures 9.3–9.8. The pressure acting behind each particular impeller is thus kept near suction pressure. Drilling several axial holes will also influence the axial thrust produced by an impeller and may affect the load on the thrust bearing set. Recall that Pressure (psi or N/mm^2) = Force/Area; hence, Force = Pressure multiplied by Area. For optimum bearing life, this force cannot be too light (skidding risk was explained in Chapter 5) or too large (resulting in insufficient oil film strength and thickness).

A heat exchanger (Figure 9.4) will add to the cost of an installation, but may be required in some services. Modern mechanical seals often include an internal pumping device – an important option that will be further explained later in this chapter.

Some applications favor the flush plan shown in Figure 9.5, although, again, its ultimate suitability for a given application or service must be assessed. The liquid injected into the seal cavity will migrate through the throat bushing and into the pumpage. This dilutes the process fluid and the injected fluid will later have to be removed by evaporation. Because evaporative processes require considerable heat the overall energy efficiency of the plant is inadvertently reduced simply because of a particular seal plan selection.

Dual seals utilize a barrier or buffer fluid to create a desirable seal environment, Figure 9.6. The cost of the seal auxiliaries shown here is clearly a factor and each application merits its own review. Connection to an external nitrogen source requires safeguards against the inadvertent backflow of barrier fluid into the

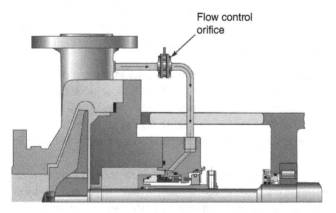

Figure 9.3 Recirculating (injecting) a product side stream (API Plan 11).

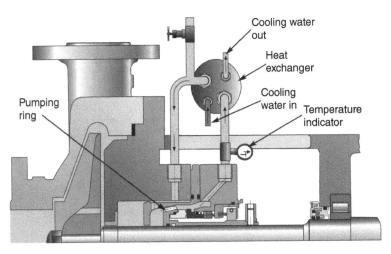

Figure 9.4 Product recirculation from seal chamber to heat exchanger and back to seal chamber per API Plan 23.

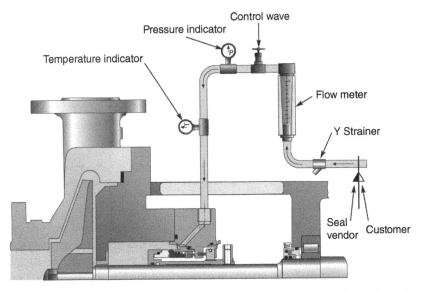

Figure 9.5 Injection of cool or clean liquid from external source into the seal chamber per API Plan 32.

external source of nitrogen. The possibility of nitrogen dissolving in the barrier fluid must be considered as well. In other words, seal selection requires study and experience.

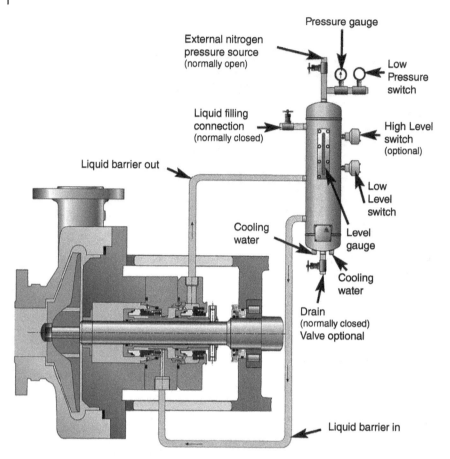

Figure 9.6 API Plan 53A – pressurized barrier fluid circulation in outboard seal of a dual seal configuration. A pumping ring maintains circulation while running; thermosiphon action is in effect at standstill.

Both Figures 9.7 and 9.8 are variations on our "go with advanced technology" theme and highlight why it is so important to have good cooperation between seal manufacturer and seal user-operator. Figure 9.8 (API Plan 62) is schematically shown again in Figure 9.9; many of these seals and flush plans have been converted to the far more effective water management approach shown in Figure 9.10.

All Flush Plans Have Advantages and Disadvantages

The many properties of fluids contacting the seal faces govern seal selection, but other considerations should be weighed as well [4,5]. Long-term reliability and savings in utilities should be given high priority. Conversely, low initial seal cost is

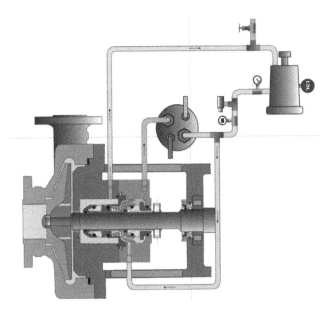

Figure 9.7 API Plan 53C – pressurized and cooled barrier fluid circulation in outboard seal of a dual seal configuration. A pumping ring (Figure 9.12) maintains circulation while running. The pressure is maintained and fluctuations are compensated in the seal circuit by a piston-type accumulator, upper right.

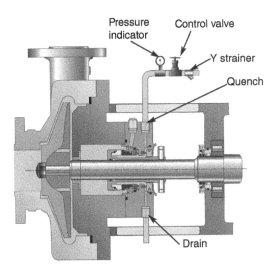

Figure 9.8 API Plan 62 – an external fluid stream is brought to the atmospheric side of the seal faces using quench and drain connections, *Q/D* in Figure 9.16. (This is a very inefficient use of water).

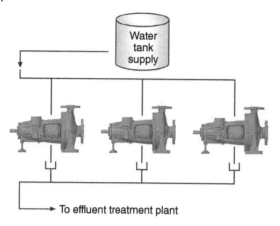

Figure 9.9 An energy-inefficient API Plan 62 in use at an older sewage treatment facility.

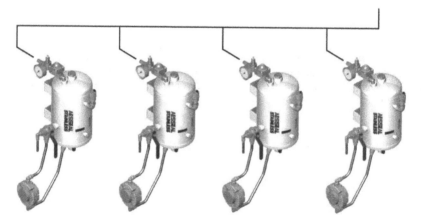

Figure 9.10 Four water management systems connect to the seal glands of four separate pumps in this sample illustration. There is no drain and no water is wasted.

rarely (if ever) a good indicator of the true value of mechanical seals and seal support systems.

Historically, Plan 23 has not received wide acceptance due to the obvious complications of applying old-style seal circulating devices or pumping rings. Utilization of modern computer-controlled manufacturing methods has helped implementing superior sealing technology.

Compact Plan 23 cartridge seals (Figure 9.11) are easily applied to both new and old pumps. These seals incorporate wide-clearance bidirectional tapered pumping devices that are far less likely to make contact with seal-internal stationary parts than older, close-clearance pumping ring configurations [6].

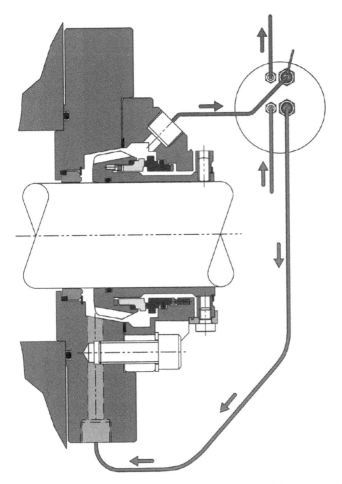

Figure 9.11 Plan 23 cross-section of a seal cavity with a bidirectional (tapered) pumping device. The flush product is pumpage that passed through the throat bushing on the left. After being cooled in a heat exchanger (upper right), this product re-enters the seal region in the vicinity of the seal faces. A bidirectional tapered pumping device (see Figure 9.12) promotes relatively high flow rates and thus cooler face temperatures.

Always Obtain the Full Picture

Seals are part of a pumping system and systems must be properly reviewed. This is demonstrated with API Plan 23, Figure 9.11. The reviewer must first recognize that the seal assembly incorporates a throat bushing which will almost (but not fully) isolate the seal chamber from the pump's case. The small

(a)

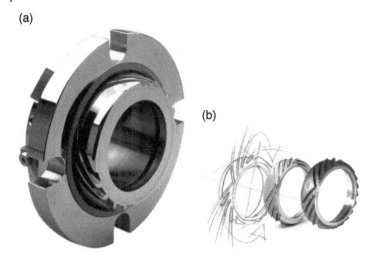

(b)

Figure 9.12 A bidirectional tapered pumping ring assembled (a) and shown separate (b).

volume of liquid in the seal chamber is circulated through a local cooler. API Plan 23 is used on hot applications and minimizes the load on the heat exchanger. It needs to cool only the heat generated by the seal faces and the heat that has migrated through the seal chamber casing. An effective pumping or circulating device (Figure 9.12) is at the very heart of Plan 23 (see also Figures 9.4, 9.16, and 9.17).

Seal Chamber Pumping Ring (Circulating Device) Technologies

Many of the pumping rings found in mechanical seals are based on a straight vane or paddle-type configuration (Figure 9.13). Typical pump-around flow rates achieved with traditional pumping rings are quite low. They will function only in the plane where the ports and the straight vanes (paddles) are located. Tangential porting will be required, and in many instances, little or no liquid is induced to flow continually over the seal faces.

Pumping screws (Figure 9.14) are considerably more efficient; however, they must rely on a dimensionally close clearance gap between screw periphery and housing bore. This close gap can be a serious liability in situations where shaft deflection or concentricity issues exist. Designs with a close gap will not comply with the API 682 (2002) requirement of a minimum radial clearance of 1.5 mm (0.060″). But screw devices with large clearance gaps (Figure 9.14) have

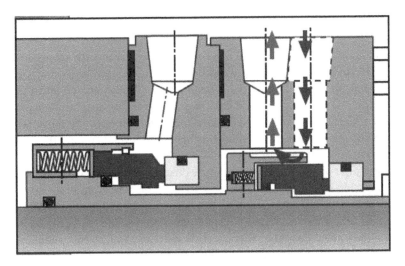

Figure 9.13 A relatively ineffective straight-vane pumping ring.

poor efficiency and can be as ineffective as the straight-vane pumping rings of Figure 9.13.

Traditional pumping screws (Figure 9.14) are, of course, unidirectional. This inherent unidirectionality leaves ample opportunity for human error on between-bearing pumps; recall that left-hand devices are required on one end and right-hand devices on the other end of the shaft.

Close radial clearances between counter-rotating surfaces can lead to component contact and galling. If a stainless steel rotary component contacts a stainless steel stationary component, galling will occur.

Some dual seals applied in industry are designed with radial clearances in the order of 0.010 and 0.020″ (0.25 and 0.5 mm). This contradicts best-practice guidelines and technical logic because pumps operating away from "Best Efficiency Point" (BEP) and most single volute pumps are known to undergo a measure of shaft deflection.

Whenever the dual seal radial clearance is less than the throat bushing clearance in the pump, a close-clearance helical screw device will be the first to make contact. All these potential issues are avoided with open-clearance bidirectional tapered pumping rings.

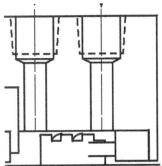

Figure 9.14 Pumping screw found in some dual seals. To be effective these auger-type helical screws will require close clearances, but close clearances introduce risk factors.

Lessons Apply to Many Services

We chose here to summarize the seal topic by using a few schematic representations. In API Plan 21 (Figure 9.15), process fluid is diverted from the discharge of the pump, sent through a restriction orifice and a seal flush cooler, and then routed into the seal chamber. The seal flush cooler is removing heat from the process stream. If the production process demands a hot fluid, such heat removal will benefit only the seal. In that case, cooling the flush stream would reduce the overall process energy efficiency.

Plan 23 can offer benefits of improved vapor pressure margin in seal chambers, thereby extending seal reliability. The reduced working temperature of Plan 23 operation has prevented coking on the atmospheric side of many mechanical seals in hydrocarbon services. This, of course, enhances seal life. There are only a few applications where Plan 21 is still preferred over Plan 23, and a competent seal manufacturer is able to point them out (Figure 9.16).

Self-contained water management systems (Figure 9.17, also Figure 9.10) are easily cost-justified in applications that must conserve water [5]. Self-checking hydraulic sensing valves are incorporated in these systems.

In essence, retrofits of water management systems are conversions from Plan 32 (Figure 9.18) to Plan 53A (Figure 9.19). These conversions often make considerable economic sense and have even improved plant output in distilleries and paper plants [2].

Upgrading to Plan 53A (Figure 9.19) was implemented on evaporator pumps at distillery units. The small pressurized tanks in Figure 9.17 contained 101 of barrier fluid and were installed at each pump. In one facility, a three-pump conversion allowed syrup production to be increased from 88 to 98 tons/hr, and the plant was subsequently able to operate fewer hours while still meeting full capacity demand.

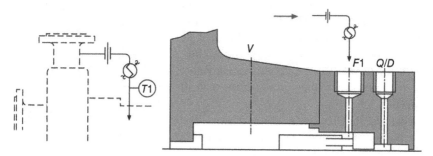

Figure 9.15 Recirculation from pump discharge through restriction orifice and cooler to seal chamber (API Plan 21).

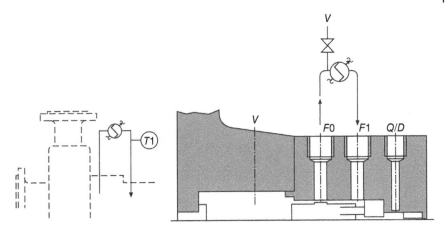

Figure 9.16 API Plan 23, recirculation from a pumping device in the seal chamber through a cooler and back to the seal chamber.

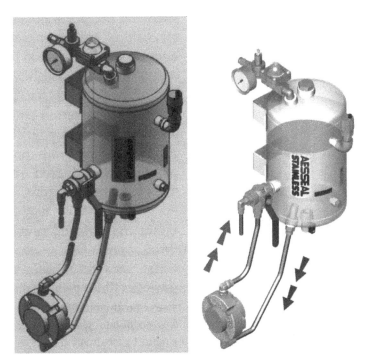

Figure 9.17 Self-contained water management systems are replacing many highly inefficient Plan 62 systems that were formerly used.

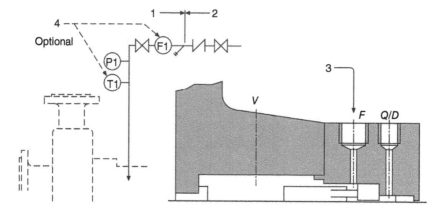

Figure 9.18 API Flush Plan 32; flush is injected into the seal chamber from an external source.

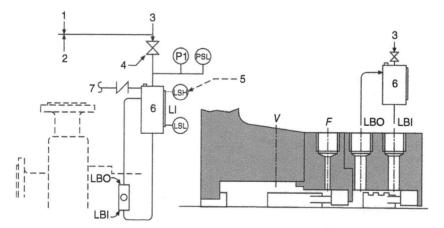

Figure 9.19 API Flush Plan 53A, where a pressurized external barrier fluid reservoir supplies clean fluid to the seal chamber at a pressure greater than that of the process (pumpage). An internal pumping device provides circulation.

What We Have Learned

- Phasing out any remaining packed stuffing boxes and upgrading to well-designed mechanical sealing alternatives makes economic sense. Follow good stuffing box packing practices until conversion to mechanical seals is complete.
- It is necessary to ascertain that vapors or gases (air) will be vented from the seal cavity. A small hole may have to be drilled in the pump cover or seal housing to accomplish this venting when "dead-ended" seal arrangements are used.

- Plan 23 is generally preferred for hot water services, particularly boiler feed water. This plan is also desirable in many hydrocarbon services, where it is necessary to cool the fluid to establish the required margin between fluid vapor pressure and seal chamber pressure.
- In just a few special cases, Plan 21 is preferred over Plan 23. Work with a competent seal manufacturer to identify these exceptions to the rule.
- Plan 53A (self-contained water management systems) are often vastly superior to inefficient Plan 32 configurations.
- Maintaining an adequate vapor pressure margin helps protect the seal faces against localized boiling of the process fluid at the seal faces. This can cause loss of seal-face lubrication and subsequent seal failure.
- Lowering the flush fluid temperature is always preferable to pressurizing the seal chamber.
- Bidirectional pumping devices are far less likely to make contact with internal seal components under conditions of pump stress. Use them whenever possible.
- Competent mechanical seal manufacturers are a valuable reliability improvement source. Make good use of their expertise and engineering know-how.

References

1 API-682 (ISO 21049); "*Pumps-Shaft Sealing Systems for Centrifugal and Rotary Pumps*", 3rd Edition, International Standard, American Petroleum Institute, September, 2004.

2 Francis, D.W., M.T. Towers, and T.C. Browne; "*Energy Cost Reduction in the Pulp and Paper Industry -- An Energy Benchmarking Perspective*", Pulp and Paper Research Institute of Canada; the Office of Energy Efficiency of Natural Resources, Canada, 2002.

3 AESSEAL plc; "*API Flush Plans, (marketing literature/laminated booklet)*", Rotherham, UK, and Rockford, TN.

4 Gurrfa, Angel, (Secretary-General, OECD); "*Water: How to Manage a Vital Resource*", OECD FORUM; 14–15 May 2007, Paris, www.oecd.org, 2007.

5 BNXS01; Carbon Emission Factors for UK Energy Use; Defra Market Transformation Program; Version 2.2, July 2007.

6 Smith, Richard; "Reconciling Requirements in API-682 Dual Seal Design Configurations", *Hydrocarbon Processing*, February 2009.

10

Pump Operation

The effects of operating issues on pump reliability and safety can be significant and must not be overlooked. Occasional reviews, comparisons, and periodic updates (or reaffirmations) of a facility's present centrifugal pump startup procedures are almost certain to reduce failure risk. Giving consideration to these periodic review activities is highly recommended.

Reliability-focused pump users view pump starting, performance monitoring, and shutting down as operator-driven reliability functions that use the following steps and sequences [1]:

Starting Centrifugal Pumps

1) Arrange for an electrician and a machinist/specialist to be present when a pump is initially commissioned. Ascertain that large motors have been checked out.
2) Close or almost close the discharge valve and fully open the suction valve. Except for axial flow pumps (pumps with high specific speed), the almost-closed discharge valve (leave it about 10% open) creates a minimum load on the driver when the pump is started.
   Assuming that the motor inrush current allows and that the motor will not kick off, the discharge valve may be just "cracked open" (again, about 10% open) before the pump is started. (A fully closed gate valve can be very difficult to open because high pressure forces the sliding valve parts into the surrounding stationary parts).
3) Ascertain that seal flush is lined-up and all related checks and procedures have been complied with.

Pump Wisdom: Essential Centrifugal Pump Knowledge for Operators and Specialists, Second Edition. Robert X. Perez and Heinz P. Bloch.
© 2022 The American Institute of Chemical Engineers, Inc. Published 2022 by John Wiley & Sons, Inc.

4) Make sure the pump is primed. Opening all valves between the product source and the pump suction should get product to the suction, but that does not always ensure that the pump is primed.

Only after ascertaining that fluid emissions are not hazardous or are routed to a safe area, open the bleeder valve at the top of the pump casing until all vapors are exhausted and a steady stream of product flows from the bleeder. It may be necessary to open the bleeder again when the pump is started, or even to shut down and again bleed off vapor if pump discharge pressure is erratic.

Note: Priming of a cold service pump may have to be preceded by "chill-down." A cold service pump is one that handles a liquid that vaporizes at ambient temperatures when under operating pressures. Chilling down of a pump is similar to priming in that a casing bleeder or vent valve is opened with the suction line open. There are three additional factors to be considered for cold service pumps:

- Chilling a pump requires time for the pump casing to reach the temperature of the suction fluid.
- The chill-down vents are *always* tied into a closed system.
- On pumps with vents on the pump case and on the discharge line, open the vent on the discharge line first for chill-down and then open the pump casing vent to ensure that the pump is primed.

Should it be necessary to have a cold service pump chilled-down and ready for a quick start, e.g. refrigerant transfer pumps during unit startup, in that case the chill-down line can be left cracked open to get circulation of the suction fluid.

5) If a minimum flow bypass line is provided, open the bypass. Be sure the minimum-flow bypass is also open on the spare pump if it starts automatically.
6) Never operate a centrifugal pump without liquid in it and never operate it with both suction and discharge valve closed.
7) Check lube oil and seal pot level (assumes dual seals).
8) Start the pump. Confirm that the pump is operating by observing the discharge pressure gauge. If the discharge pressure does not rise, stop the pump immediately and determine the cause.
9) Open the discharge valve slowly while watching the pressure gauge. The discharge pressure will probably drop somewhat, then level off and remain steady. If it does not drop at all, there is probably a valve closed somewhere in the discharge line. In that case, open the discharge valve. Do not continue operation for more than a few seconds with a closed discharge valve or a blocked line.
10) If the discharge pressure drops to zero or fluctuates widely, the pump is not primed. Close the discharge valve and – if safe – again open the bleeder from the casing to exhaust vapor. If the pump does not pick up at once, as shown

by a steady stream of product from the bleeder and steady discharge pressure, shut down the pump and check for closed valves in the suction line. A dry-running pump will rapidly destroy itself.

11) Carefully check the pump for abnormal noise, vibration (using vibration meter) or other unusual operating conditions. An electrician and machinery engineer should be present when pumps are started up for the very first time, i.e. upon being initially commissioned or after serious repairs.

12) Be careful not to allow the bearings to overheat. Recheck all lube oil levels.

13) Observe whether or not the pump seal or stuffing box is leaking.

14) Check the pump nozzle connections and piping for leaks.

15) When steady pumping has been established, close the startup bypass and chill-down line (if provided) and check that block valves in minimum flow bypass line (if provided) are open.

16) Re-check lubricant quantity (level) and bearing housing temperatures. If too hot to the touch (the human hand does not tolerate exposure to more than 170 °F or 76 °C for over five seconds), ask for an exact measurement using infrared or surface pyrometer.

Surveillance of Pump Operation

1) Especially not only at start of pumping but also on periodic checks, note any abnormal noises and vibration. If excessive, shut down.

2) Note any unusual drop or rise in discharge pressure. Some discharge pressure drops may be considered normal. When a line contains heavy, cold product, and the tank being pumped out contains a lighter or warmer stock, the discharge pressure will drop when the volume in the line has been displaced.
Also, discharge pressure will drop slowly and steadily to a certain point as the tank level is lowered. Any other changes in discharge pressure while pumping should be investigated. If not explainable under good operating conditions, shut down and investigate thoroughly. Do not start up again until the source of the trouble has been found and remedied.

3) If lubricated by oil mist, periodically check oil-mist bottom-drain sight glass for coalesced oil or water and drain, if necessary.

4) On open (old style) oil-mist systems, regularly check both oil level and stray oil-mist flow from vents or labyrinths.

5) Seal of oil pots need to be checked regularly for correct level. Refill with fresh sealing liquid (a special oil, propylene glycol, or methanol, as specifically required and approved for the particular application).

6) Periodically check for excessive packing leaks, mechanical seal leaks, or other abnormal losses. Also, check for overheating of packing or bearings. When in

doubt, do not trust your hand. Use an infrared gun or a surface pyrometer instead. Note that excessive heat may cause bearing failure and may result in costly and hazardous fires.

7) When a pump has been repaired, place it in service as soon as possible to check its correct operation. Arrange for a machinist (and electrician after extensive overhaul) to be present when the pump is started up.

Optimum Pump Switching Interval

It is common to have twin pump installations in *critical* pumping services, where one pump is in service and the second pump is idle. A high level of reliability is achieved by always having an installed spare ready to go in the event the pump in service fails unexpectantly. Some sites refer to these twin pumps as "A & B" pumps, while others refer to them as "main and spare" pumps. These pumps are normally installed next to each other (see Figure 10.1) and are tied into common suction and discharge headers. If the pumps are installed properly, similar operating performance can be expected, regardless of which pump is in service. Rapid switching between these twin pumps by capable operators can ensure that the process flow is never interrupted long enough to result in a production upset or outage.

However, the key to twin pump reliability is exercising both pumps regularly to ensure their serviceability. "Best Practices Plants" have determined monthly switching of the critical "A" and "B" (also referred to as main and spare) pumps to be the most appropriate (optimum) and cost-effective pump-switching strategy.

Figure 10.1   A typical twin pump installation. Notice both pumps are tied into common suction and discharge headers.

This optimum pump-switching interval can prevent bearings and mechanical seals from seizing in idle pumps. Furthermore, regular pump switching allows the mechanical condition of both pumps to be evaluated periodically under normal operating conditions, i.e. vibration data can be collected, bearing temperature can be assessed, etc.

Centrifugal Pump Shutdown

Again, reliability-focused pump users consider it worthwhile to occasionally reassess their shutdown procedures. These are among the many operator-driven reliability functions, where the three primary job functions (i.e. operations, maintenance, project-technical) merge, and where there needs to be a consensus among these three job functions.

Reliability-focused pump users consider using the following procedures:
Special note: On hot service pumps and upon shutting down, commission ("line up") the "re-start liquid fill" piping. Make sure the mechanical seal region is not exposed to viscous pumpage:

1) Close the discharge valve or leave open about 5% to facilitate opening fully. This takes the load off the motor and may also prevent reverse-flow through the pump.
2) Shut down the driver.
3) If the pump is to be removed for mechanical work, close the suction valve and open the vent lines to flare or drain, as provided. Otherwise, leave suction valve open to maintain pump at the correct operating temperature.
4) Shut off steam tracing, if any. Continue oil-mist lubrication, if provided. Oil mist protects bearings and prevents intrusion of atmospheric contaminants during periods of nonrunning.
5) Shut off cooling water (which is provided on sleeve bearings) and sealing oil, etc., if the pump is to be removed for mechanical work.
6) At times, an emergency shutdown may be called for. If you cannot reach the regular starter station (in case of fire, for example), stop the pump from the remote starter box, which is located some distance away and is usually accessible. If neither the starter station nor the remote starter box can be reached, call the electricians.

Still, do not – as a part of regular operations – stop pumps from the remote starter box. Use the regular starter station instead.

Please note that these procedures are of a general nature and may have to be modified for nonroutine services. Always review pertinent process data, as applicable. Rewrite these instructions in concise sentences if they are to become part of checklists that operators are asked to have on their person while on duty.

Addressing Post Start-up Issues

Once a pump has been started up and warmed up to its normal operating temperature, it is the operator's responsibility to continue to assess its mechanical condition until deemed reliable. To this end, operators must visit and observe recently commissioned or repaired pumps more frequently. During these visits, the operator should walk around the pump, look for leaks, and listen for unusual sounds. As part of these visits, the operator must determine if

1) The pump flow is normal and steady
2) Pump suction and discharge pressures are normal and stable
3) The pump speed is normal and steady
4) Pump vibration and sound levels seem normal

Here is some guidance to help operators if they find that a pump is not functioning properly or showing signs of distress. (Please note that advanced troubleshooting is beyond the scope of this chapter, so the following advice will touch on the most common field problems and how to approach them.)

Probably the most common issue encountered after a pump start-up is that the net flow is lower than expected. Here are some issues to look for if a low flow condition is encountered:

1) Make sure the suction and discharge pump valves are fully open. A pinched valve will cause a restriction that will affect pump flow.
2) Check for open bypass valves: It is a very common problem that a normally closed valve on the discharge header is accidentally left open during the start-up process. While the overall pump flow may be normal, the net pump flow will be lower than expected if some of the flow is diverted to another location.
3) Another possible cause of a low-flow condition is a miscalibrated minimum flow spillback system. This would result in spillback flow when it is not required. Check the position of the minimum flow valve and its bypass to ensure they are working as they should. On applications with a main and installed spare, a leaking check valve on the idle pump will act like an open bypass line. When this occurs, some of the discharge flow will exit from the discharge header, through the leaking check valve, out the idle pump's suction, and back to the suction tank or vessel, thereby reducing the net flow. This issue can be temporarily managed by closing the discharge valve on the idle pump until the check valve can be removed and repaired.
4) *Check for a leaking safety relief valve on the pump header or discharge line*: A leaking safety relief valve will act like an open bypass line, which will reduce

the net forward pump flow. An ultrasonic leak detector can be used to identify a leaking relief valve.

5) *Check the driver's speed*: A pump's speed is critical to its performance. We know that the heads and resulting flows generated by centrifugal pumps are a strong function of speed.

Let us quickly review the pump affinity laws, which are a set of mathematical relationships that predict how capacity, head, and horsepower are affected by changes in impeller diameter (D) and shaft speed (N) (see Table 10.1). We can see from studying the table that flow (Q) is directly proportional to speed and that head (H) is proportional to speed squared. Therefore, if operating speed falls below the design speed, then the flow will also drop below the design flow accordingly, assuming there is not some other system issue. For example, a 5% drop is pump speed will result in a $1 - (0.95)^2$ or >9% loss in head than expected, which, depending on the shape of the pumping system curve, could lead to up to a 5% loss of flow.

There are several possible causes of low-speed condition:

- *The wrong electric motor was installed*: It is not unheard of to see an 1800 rpm motors installed, where a 3600-rpm motor was needed. Mistakes can be made when attempting to use a surplus motor and not double-checking its rated speed. If a low-flow condition occurs after a motor changeout, check the rotational speed first to ensure you are operating at the design speed.
- *The variable frequency drive (VFD) was not been calibrated properly*: It is vital to verify that a 100% flow requirement will deliver a 60 Hz command from the VFD to the electric motor. There was an actual case where a 100% flow signal resulted in a 58 Hz command to the VFD, which resulted in a fixed reduction in available head pressure from the pump.

Table 10.1 Pump affinity laws.

Diameter change only	Speed change only	Diameter and speed change
$Q_2 = Q_1\left(\dfrac{D_2}{D_1}\right)$	$Q_2 = Q_1\left(\dfrac{N_2}{N_1}\right)$	$Q_2 = Q_1\left(\dfrac{D_2}{D_1} \times \dfrac{N_2}{N_1}\right)$
$H_2 = H_1\left(\dfrac{D_2}{D_1}\right)^2$	$H_2 = H_1\left(\dfrac{N_2}{N_1}\right)^2$	$H_2 = H_1\left(\dfrac{D_2}{D_1} \times \dfrac{N_2}{N_1}\right)^2$
$BHP_2 = BHP_1\left(\dfrac{D_2}{D_1}\right)^3$	$BHP_2 = BHP_1\left(\dfrac{N_2}{N_1}\right)^3$	$BHP_2 = BHP_1\left(\dfrac{D_2}{D_1} \times \dfrac{N_2}{N_1}\right)^3$

- *The steam turbine speed is incorrectly set, usually too low*: Check the pump speed and attempt to adjust the steam turbine speed to the correct speed setting.
- *The steam turbine speed cannot reach required speed*: This issue could be the result of either a plugged steam inlet strainer, malfunctioning governor, or a worn-out steam turbine. Have a qualified mechanic inspect the steam turbine to determine the best and safest path forward.

6) Check the pump's suction conditions. Suction problems commonly result in flow problem, so start by checking the pump's suction pressure and compare it to what is normal. The suction pressure should be steady and close to the design value. If the suction pressure is too low or erratic, you should suspect either a plugged suction strainer, a loss of suction level, or cavitation. First check to see that the suction level is normal before considering removing and inspecting the suction strainer. Best-in-class (BiCs) use temporary suction strainers to catch construction debris during initial commissioning. They will not leave these strainers in place.

7) *Check current process conditions, i.e. density, viscosity, vapor pressure, etc.*: After all the simple operating factors have been checked, then it is time to consider that operating conditions may have deviated from design values. Fluid properties can deviate whenever the composition of the fluid being pumped changes. There may be occasions when fluid composition during startup may be quite different than the composition during normal operating conditions.

There are many possible causes of pump flow problems, but in the interest of time, we recommend that site personnel approach field problems in this order:

a) First, check the pump speed using a portable tachometer or by reading a dedicated tachometer. This is easy to check.
b) Then, check for open bypass valves and leaking check valves in services with installed spares.
c) Next, check the pump's suction conditions.
d) Lastly, check fluid properties, i.e. density, viscosity, vapor pressure, etc. Since these are normally not issues, they should be the last items to check.

Other commonly encountered problems after start-up are abnormally high vibration levels and abnormal pressure pulsations. Abnormally high vibration levels after a startup can be caused by several factors. The most common vibration related issues are the following:

1) *A foreign object gets lodged in a pump impeller*, which is generally characterized by high, running speed (1×) vibration components in the vibration spectra. (a) After process upsets, tower bottoms pumps are especially prone to

plugging with tower hardware or debris if they get dislodged and fall to the bottom of the tower. (b) After unit startups, any trash or debris left inside process piping can find its way into pump suction lines and eventually get stuck in impellers.

2) *Misalignment between a pump and its driver*, which can show up as high, one and two times, running speed (1× and 2×) components within the vibration spectra.

3) *Internal looseness of pump components*, which typically generates vibration spectra with multiples of running speed components, i.e. 1×, 2×, 3×, etc. The presence of high 3× components in the pump spectra can be a strong indication internal looseness.

4) *Cavitation*, which tends to generate random, higher frequency, broadband vibratory energy that often shows up as blade pass frequency harmonics (f_{vp} = number of impeller vanes × speed in Hz) superimposed within the vibration spectra. (Typically, when a pump is cavitating, you will hear a sound similar to gravel traveling through the pump and notice an unsteady pump discharge pressure.)

5) *Off design operation,* which usually generates flow-related vibration components that are generally multiples of the vane pass frequency, i.e. f_{vp}, $2 \times f_{vp}$, $3 \times f_{vp}$, etc. Start the analysis by checking to the see where the pump is operating on its performance curve. Operating too far away from a pump's best efficiency point (BEP) can result in internal recirculation, which will lead to vibration and cavitation-like symptoms. An easy test to try is to pinch the pump's discharge valve to reduce flow. If the pump quiets down, then the problem is probably cavitation, but if the pump gets noisier, then the problem may be related to operating too far away from the BEP.

Abnormal pump pressure pulsations can be caused by

1) *Cavitation*: If cavitation has not been seen before, then there is probably some type of flow restriction dropping the suction pressure below the liquid's vapor pressure. The most likely source of a restriction is a partially closed suction valve or a plugged pump suction strainer.

2) *Loss of adequate fluid level*: A common problem seen is that a tank or vessel level is either completely lost or well below the recommended level. As the liquid level drops, vapors can begin to get entrained in the fluid going to the pump, which can lead to cavitation-like performance. Ensure that the liquid level is restored so that there is adequate retention time to allow vapors to disengage. In severe cases, a low liquid level can lead to vortexing, which allows vapor to get pulled directly into the fluid stream going into the pump.

3) *Off-design operations*: See the comments above about the effects of off-design operations and how to determine if this is a problem.

Avoiding Parallel Pump Operating Problems

There are times when two or more centrifugal pumps must be installed side by side so that their flows combined. This design configuration, which is referred to as a parallel pump installation (see Figure 10.2), is used to provide operating flexibility by enabling Operating Personnel to run one or more pumps, depending on process demand. However, the indiscriminate parallel operation of centrifugal pumps can lead to significant flow imbalances, which can result in pump vibration, pulsations, overheating, and even failure. Pump users should always check to see that pumps are similar in performance capabilities before attempting to operate them in parallel. To be hydraulically similar, every pump placed in parallel service should be capable of generating the same head at the required flow. If any pump is not capable of developing the same pressure as its sibling pumps, there is a possibility that it will not be capable of opening its own check valve. As a consequence, it will operate in a zero-flow condition, which would likely result in rapid and catastrophic failure.

As added insurance, individual flow meters may be used to ensure pumps are operating at a safe flow when operating in a parallel arrangement. However, users rarely have the luxury of having a dedicated flowmeter on every pump. It is more common to have a total flowmeter for a group of pumps. In multiple pump installations, project groups will rarely provide a flowmeter for each pump due to cost, lack of real estate, or lack of economic justification.

It is common for project managers to believe that similar pump pairs are identical, and that parallel operation will always result in balanced pump flows. However, in the real world, there is always the possibility that one or both pumps

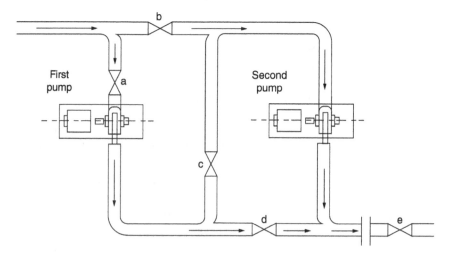

Figure 10.2 Two centrifugal pumps in parallel operation.

will operate in an unsafe condition due to internal wear, slight differences in pump construction, or due to assembly issues. For this reason alone, it would be prudent to have some type of individual pump flow or load measurement installed if pump sets are to be operated in parallel. If a flowmeter is not possible, then a power meter on each pump can be used to monitor and balance pump flows.

For larger, critical pump consider flow meters with individual flow controllers to ensure balanced flows. In more complex arrangements with wide flow range requirements, a PLC (programmable logic controller) capable of shutting down and starting pumps as required by the process may make sense. In rare cases, spillback controls may also be required to protect pumps required to perform over a wide flow range.

Additional Start-up and Monitoring Tips

1) When starting up a centrifugal pump during its commissioning or after a unit outage, it is likely that the pump's discharge line will be empty and unpressurized, which can result in an excessive flow immediately after the pump reaches rated speed [1]. A situation of low or zero back pressure can result in a temporary operating condition called "runout." During "runout "events, the pump operates toward the high flow end of the pump curve, which can result in a higher than normal amp load (for low suction-specific speed pumps) and high vibration levels due to off-design operation. The avoidance of "runout" is the reason why extra care should be taken to ensure the discharge valve is in the partially open position, i.e. about 10% open or less, until back pressure is fully developed and the discharge valve can be fully opened. "Runout" operation for even a short time can result in severe pump and seal damage. In severe cases, "runout" operation can even lead to metallurgical fatigue on small bore piping, i.e. less than 2″ diameter, and fittings due to wildly erratic swings in flow.

2) Vertical turbine pit pumps in open pits are susceptible to air entrainment due to vortexing caused by low-pit levels and plugging due to trash build-up in the pit. Operators should remain vigilant for the occurrence of these two common problems. Low or erratic flows and pressures are good indicators of abnormal suction conditions. If plugging due to trash is a frequent occurrence, pit covers and secondary screens should be considered.

3) In the flow region around the BEP, most vertical turbine pumps are designed so that the hydraulic upthrust forces acting on the rotor are small compared to the magnitude of the hydraulic downthrust forces. (Upthrust rotor forces are associated with the momentum changes in the liquid flow passing through the impellers.) However, at high capacities, i.e. end of curve flows, the sum of the rotor upthrust forces can exceed the sum of the downthrust forces, resulting in a net upthrust load on the rotor. A significant net upthrust hydraulic rotor force can destabilize the rotor and result in excessive rotor vibration levels,

which can damage seals and bearings and, in severe cases, lead to a permanently bent shaft [2]. Check with the pump manufacturer to see if any upthrust conditions are expected and design your thrust bearings accordingly. Start-Up Considerations: When a vertical turbine pump is first started with a fully open discharge valve, it can temporarily operate at a high flowrate. This occurs because the motor will reach rated speed in a few seconds, but it may take somewhat longer for the pump head to build up enough to move the flow back into a safe range. The pump could therefore momentarily operate in the flow range where a net upthrust may occur. If an upthrust condition is possible during start-ups, then the motor thrust bearings must be designed to reliably handle these possible momentary upthrust conditions. Additionally, operating procedures should be written to require a partially closed discharge valve during startups to prevent or mitigate monetary upthrust events due to unwanted high flow conditions.

4) It is good practice to preheat hot service pumps before starting them. A major concern for pumps using wear rings with galling tendencies is that of rubbing between the rotating and stationary wear rings. Rubbing can be caused by either pump casing distortion or differential expansion between the rotor and casing. Users don't have any control over casing distortion or poor piping design, but they can control the differential growth between the rotor and stator by preheating their hot pumps.

One common means of preheating a pump in hot applications is by installing a small bypass (1″ or smaller as required) around the pump's discharge and check valve (see Figure 10.3). For larger pumps, a better practice is to install warm-up piping that allows hot liquid to flow into the bottom of the casing and out the discharge nozzle to ensure the entire pump is adequately preheated. Multiple warm-up lines are recommended at the bottom of large multistage

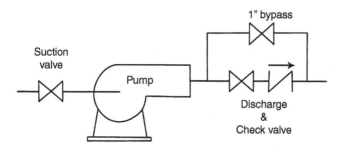

Figure 10.3 A hot pump with warm-up piping that allows liquid to bypass the discharge valve and check valve when it is not in service. Hot liquid flows from the discharge header back, through the bypass line and pump, and then back to the suction header. *Note*: A ¾″ bypass valve or a ¾″ valve with a flow restriction orifice can be used to control spillback flows on smaller centrifugal pump installations.

pump casings to evenly distribute internal flow to ensure thorough preheating. Do not forget to safeguard the mechanical seal's temperature environment. To prevent damaging the seal, mere preheating prior to startup may not be sufficient; protection of a seal's bellows component against twisting in semisolid liquid environments will be needed. Therefore, give consideration to keeping the seal region heated at all times.

Any pump with galling tendencies that operates above 150 °F should be preheated prior to start-up. The risk of galling or grabbing is proportional to the pump's operating temperature. So the hotter the pump operates the more care should be taken to ensure it is properly preheated. Insulation should be used to assist in the warm-up process and to maximize the thermal growth of the casing. Prior to start-up an infrared temperature gun or contact thermometer can be used to ascertain the casing is properly heated.

Soon after idling pumps handling products that tend to solidify, crystallize, or clog, it is good practice is to flush their casings (and seal cavities) with a clean, low viscosity fluid that can easily be pumped out during the next start-up. Furthermore, any pump handling potentially troublesome or hazardous fluids should be thoroughly flushed before it is removed and sent out for repair. If the fluid being handled is toxic or caustic, a casing decontamination step may be required to ensure it can be safely transported and then handled by the receiving repair shop.

To efficiently flush out a pump casing, a suitable flushing medium and well-designed ingress and egress piping are required. Adequate fluid velocities must be created inside the pump to properly sweep out most of the fluid left inside after it has been shut down. Operations and pump piping designers must work together to guarantee pump flush piping is properly configured and installed. A detailed procedure should be written and followed to ascertain the pump casing and seal cavity, or cavities, are thoroughly flushed out during the entire process.

Here is a listing of helpful tools for monitoring centrifugal pumps. Every control room should have access to the following:

a) *Vibration meter:* A handheld vibration meter should be kept in control rooms to allow operators to assess the operational fitness of pumps that are suspected of having a vibration issue. Even if there is a vibration monitoring program on site, it still makes sense to have a portable vibration meter to double-check the mechanical condition of pumps in the field. Every operator should be trained to use a vibration meter and given straight forward "go" and "no-go" criteria for determining when to call on an expert to investigate the issue.

b) *Tachometer:* As mentioned earlier, it is important to have a means of determining the pump speed when flow problems are encountered.

c) *Temperature gun or contact thermometer:* Portable thermometers can be used to monitor bearing housing and oil reservoir temperatures. These devices can also be used to spot check mechanical seal problems by monitoring the temperatures of seal inlet and outlet lines.

d) *Replacement pressure gages*: Field pressure gauges should be kept in working order. Vital pumps should always have working suction and pressure gauges to allow for pump performance monitoring.

e) Detailed shutdown and startup procedures along with mechanical seal system line-up instructions for all pumps should always be available in every control room.

Capabilities that every production site should have access to

a) *Vibration analysis capability*: Every site should have the ability to perform vibration spectral analyses when required. When vibration problems arise, the ability to quickly analyze the vibratory content will enable troubleshooters to quickly focus in on the root cause of the vibration and act accordingly.

b) *Amp meter or power meter*: Pump power is a useful measurement when troubleshooting field problems; it serves to determine if the pump is properly loaded or not. Overloading will result in unwanted motor trips. Power meters are preferable to amp meters; power meters are more accurate at lower power levels.

c) *Oil analysis capability*: Oil analysis capabilities provide a means of determining oil condition, tracking contaminants, and detecting early signs of bearing wear. However, to be successful an oil analysis monitoring program should have a database of past analyses. This will allow a comparison of current lab results with past results and allow trends to be established and analyzed.

d) *Infrared thermography (IRT) equipment*: Experience shows that temperature is an excellent way to monitor the performance and condition of pumps. Temperature measurements are used to diagnose developing mechanical problems. Thermography is useful to identify and analyze thermal anomalies around centrifugal pumps, such as lubrication issues, misalignment, wearing components, abnormal load conditions. IRT is based on measuring the distribution of radiant thermal energy (heat) emitted from a target surface and converting this to a surface temperature map or thermogram. Thermal energy may be in the form of frictional losses within machines, energy losses within machines, fluid characteristics, or a combination of all these factors. IRT can provide complete thermal images of a centrifugal pump without the need for physical attachment, while requiring little setup and providing visual results quickly.

What We Have Learned

- Generalized pump startup and shutdown procedures apply to the majority of radial impeller centrifugal process pumps.
- Specialized procedures are needed for mixed flow and axial flow pumps.

- Precooling and/or preheating impose special requirements on cold and hot service pumps.
- Venting requirements can differ from service to service, making it important to view pumps as part of a system.
- The operators are the eyes and ears of the plant. They are trained to be the "first responders" and must accept the responsibility to be the first to notice deviations from the norm.
- The operators' job must include data collection and equipment surveillance. Once they spot a deviation from normal pump behavior, they are empowered to ask other job functions within the company to assist in analyzing the root cause.

References

1 Perez, R. X.; *"Operator's Guide to Centrifugal Pumps (Volumes 1 and 2)"*, Xlibris Publishing Co. (Vol. 1) and 2015 Vol. 2), 2008.
2 Smith, D. R, and Price, S.; "Upthrust Problems on Multistage Vertical Turbine Pumps", Proceeding of the Twenty-Second International Pump Users Symposium, pp 46–57, 2005.

11

Impeller Modifications and Pump Maintenance

Please recall that this text is not intended as a pump maintenance book. Many pump manufacturers have compiled recommendations on pump maintenance, and these vary from one manufacturer to the next. They generally differ among pump types and models. Yet, there are some common threads or essentials to be considered.

Maintenance Essentials

Maintenance is to prevent deterioration and, if necessary, restore things to the as-sold or as-installed state. Routine preventive maintenance of process pumps is generally limited to oil replacement and replenishing. Next in line is mechanical seal maintenance, including seal replacement.

Chapters 6 and 7 dealt with lubricant application, cooling, and lube types. Lubricant replacement (oil change) is done four times per year at a particular US refinery – one that clings to the very outdated and wasteful tradition of using cooling water on its pumps. Another refinery with soundly engineered constant level lubricators, advanced bearing housing protector seals, and premium quality synthetic lubricants (ISO VG 32 and 68) changes oil every three years.

Of course, oil change issues do not exist at facilities that use pure oil mist instead of liquid oil. Some pump manufacturers decline to comment on the fact that no maintenance is needed for oil-mist lubricated bearings. Declining to comment may simply be the result of not having updated one's knowledge base.

After a repair has been completed, proper installation and its verification are always needed. But additional maintenance tasks vary from plant to plant. These include condition monitoring by periodically comparing head vs. flow data – an

Pump Wisdom: Essential Centrifugal Pump Knowledge for Operators and Specialists, Second Edition. Robert X. Perez and Heinz P. Bloch.
© 2022 The American Institute of Chemical Engineers, Inc. Published 2022 by John Wiley & Sons, Inc.

operator-technical function – and spotting deviations before they give rise to process-operational and safety issues.

It is always good to remember that a single deviation rarely causes pumps to fail. Regrettably, people assume that they can get away with another deviation, and a third one, and a fourth one. After a while, living with these deviations becomes the "new normal." Then, suddenly, just one more deviation occurs. Now the pump fails massively and puts human lives and physical assets at great risk. The hunt for culpable parties begins, and the legal profession is mobilized. Interviews and depositions are being arranged. Orders of magnitude more money is spent than what it would have cost to avoid the problem in the first place. How, then could the problem have been avoided? By superior maintenance!

Superior Maintenance Requires Upgrading

Suppose there was a design flaw, or some other hidden, elusive issue that is responsible for causing repeat failures, short run lengths of a pump, or whatever else. More than traditional maintenance effort may be needed, which gets us to the subject of superior maintenance. Superior maintenance is a reliability improvement task. The best-performing or Best-in-Class companies are not repair-focused; they are reliability focused. These are companies that have given some of their personnel the task of determining if upgrading to a better component is feasible.

If an upgrade is feasible, the contributor is asked to calculate its cost. A best-performing company would then ask what this upgrade effort will be worth, perhaps in a simple benefit-to-cost or payback estimate.

A good example involving upgrade issues would be Figure 11.1. The "cooling fan" in Figure 11.1a is simply too small to be of any value. The well-dimensioned fan shown in Figure 11.1b will probably be very effective. Recall, however, that our earlier Chapter 6 highlighted the potential drawbacks of applying cooling only to bearing outer rings. The hot inner rings may thermally expand and cause bearing preload increases that will then reduce bearing life.

Suppose an effective fan was used. In that case, the entire design must be reviewed for pressure differences surrounding the bearings. These differences might affect lube oil flow or oil-level conditions in the adjacent bearing set. Therefore, many bearing housings must be upgraded to reduce failure risk. In view of the discussion in Chapter 6 and comments on the deletion of bearing cooling, the best course of action would be to discard both of the two fans in Figure 11.1. Also, it would be prudent to

a) implement means of pressure equalization across the thrust bearing set
b) use an appropriate synthetic oil

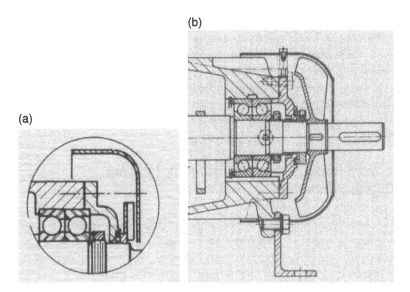

Figure 11.1 Ineffective "cooling fan" (a) and a reasonably effective fan (b).

c) consider the various shortcomings of using oil ring(s) shown in Figure 11.1b
d) question and then verify the axial load capacity of the snap ring in the housing bore shown in Figure 11.1b.

Impeller Upgrading with Inducers

Three of the many thousands of impeller configurations and geometries available are illustrated in Figure 11.2. Closed and open impellers are shown in Figures 11.2a and 11.2c; an inducer-type closed impeller is shown in Figure 11.2b.

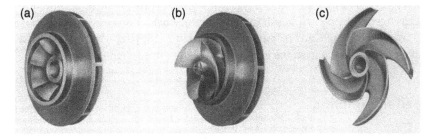

Figure 11.2 Closed impeller (a), closed impeller with inducer (b), open impeller (c).

Inducers are often custom-designed for a particular pump application. They lower the NPSHr of an impeller, but do so only in the vicinity of best-efficiency flow (BEP). If pumps with inducer-equipped impellers are operated at flows substantially above or below BEP, their NPSHr may actually increase compared to that of a standard impeller not so equipped. Instead of the continually increasing slope of the NPSHr curve shown earlier in Figure 1.4, the NPSHr curve associated with an inducer would probably have a distinct U-shape.

Distance from Impeller Tip to Stationary Internal Casing Components

Any particular process pump is designed with a casing which, of course, surrounds the pump rotor. For this rotor and its impeller to turn freely while, at the same time, accommodating a certain amount of shaft run-out and shaft deflection (see earlier Figure 1.9) there needs to be a rather liberal clearance.

In practice, one makes a distinction between two different gaps shown in Figure 11.3: Gap "A" and Gap "B." Both of these are depicted in Figure 11.3b. Gap "A" is the distance from the impeller disc and cover (the cover is sometimes called a shroud) peripheries to the nearest casing or other stationary part. That part is usually the tongue, or cutwater. Some pumps are designed with casing internals called diffuser-style or vaned, others are simply an unimpeded stationary passageway for the fluid leaving the impeller. Regardless of casing-internal construction, a gap of 50 mils (0.050 in. or 1.2 mm) is chosen for best possible efficiency and so as to minimize leakage-induced flow turbulence.

Gap "B" is the average distance from the impeller vane tips (the diameter D) to the casing-internal stationary part where the fluid coming off the impeller vane tips enters a stationary passageway. This part of the stationary passageway is usually called the cutwater, or volute tongue. Recommended (radial) minimum, maximum, and preferred values for Gap "B" are given, in Table 11.1, as a percentage of the impeller radius. In large pumps (typically pumps with drivers in the over 250 kW (335 hp) range), too small a gap will often increase vibration amplitudes, whereas too large a gap will result in a loss of efficiency.

As an alternative to Gap "B" modifications, volute chipping – the removal of a small amount of metal from the tongue or cutwater – is sometimes feasible [1].

Impeller Trimming

Virtually every textbook on pump engineering contains guidelines on impeller diameter reductions ("trimming") needed to permanently affect the rate of

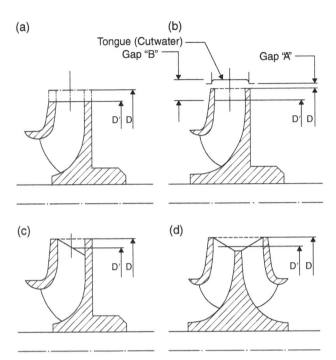

Figure 11.3 Gap designations and impeller trim methods [1], including the following: (a) Cut across entire outside diameter. (b) Trimmed vane tips; disc and cover O.D. remain untouched, i.e. as originally provided by vendor. (c) Oblique cut on a single-flow impeller. (d) Oblique cuts (mirror-image like) on a double-flow impeller. *Source:* Dufour and Nelson [1].

Table 11.1 Recommended radial gaps for process pumps.

| Type | Gap "A" | Gap "B" percentage of impeller radius | | |
		Minimum (%)	Preferred (%)	Maximum (%)
Diffuser	50 mils	4	6	12
Volute	50 mils	6	10	12

flow and the head created by a constant-speed centrifugal pump. Yet, many of these texts lead to the erroneous assumption that suitable diameter reductions are made by simply machining uniformly across the full periphery – impeller cover (shroud), vane tips, and disc. These texts use Euler's fan law equation, a mathematical relationship in which n, D, Q, H, and P are, respectively, rotor (impeller) rpm, impeller diameter, flow, head, and power demand. In this equation, the subscript 1 refers to the original value and subscript 2 indicates the new value [2].

$$\frac{Q_1}{Q_2} = \frac{n_1 D_1}{n_2 D_2}$$

$$\frac{H_1}{H_2} = \frac{n_1^2 D_1^2}{n_2^2 D_2^2}$$

$$\frac{P_1}{P_2} = \frac{n_1^3 D_1^3}{n_2^3 D_2^3}$$

However, the various assumptions on which these relationships are based are rarely giving precise answers. Flow angles and the resulting velocity relationships are being disturbed by trimming. Experience shows that in real-world situations, a reduction of impeller diameter greater than 15% of the original full diameter should not be allowed. To be on the safe side and so as not to cut too much, a prudent pump specialist will look at the mathematically derived diameter reduction (sometimes called the "fan law") and then make a trim cut of only 70% of what the Euler-based fan law math requires.

Suppose, as an example and with no speed change, we wanted a pump head reduction from previously 680 ft to a new head of 580 ft. Suppose also we had an original diameter D_1 of 13.00 in. and wished to determine the new diameter D_2. After doing the algebraic transposing and square root extracting, we quickly find a new diameter of 12.14 in. – 93% of the original. Using the 70% rule, we would now remove not $13.00 - 12.14 = 0.86$ in. but, instead, only $(0.7)(0.86) = 0.6$ in. In other words, field experience tells us the impeller should be trimmed to 12.4 in. It may not be the exact needed diameter, but we will have avoided the risk of trimming off too much metal.

Good practice would be to trim only the vane tips (Figure 11.3b) and to leave both the impeller cover (sometimes called impeller shroud) and disc at the recommended Gap "A" diameters. Best practice would be to cut obliquely (see Figure 11.3c) for greatest structural support and to ward off resonant vibration. Vibration of unsupported regions can lead to cracks and failure.

With oblique cuts, the average diameter D' is used as the relevant diameter D_2 in the Euler equation. Either oblique cutting or partial removal of unsupported regions (called "scalloping" in Figure 11.6) is done to reduce the risk of fatigue failure.

Impeller Wear Rings

Modern process pump impellers are commonly fitted with plain wear rings. Wear rings separate regions of high pressure (discharge) from regions of low pressure

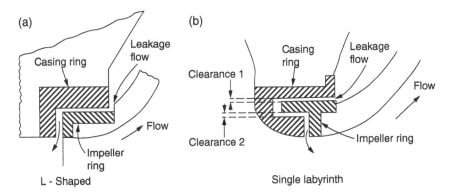

Figure 11.4 L-shaped (a) and labyrinth-shaped impeller wear rings (b). *Source:* Karassik et al. [2].

(suction). Wear rings are considered replaceable parts and the gap between stationary (casing-mounted) and rotating (impeller-mounted) wear rings should follow the guidance found in API-610.

Most wear rings are plain cylindrical, although step and labyrinth-types (Figure 11.4) are occasionally used in efforts to reduce leakage flow. Their effectiveness does not appreciably differ from that of a plain wear ring. The time and effort needed to gain marginal improvement by implementing clearances 1 and 2 (Figure 11.4b) are rarely worth it. However, selecting a carbon-reinforced perfluoro-alkoxy (Vespel® CR-6100) wear ring material will be well worth considering.

Unlike the plain wear ring set shown in Figure 11.5a, the wear rings with enhanced contours, Figure 11.5b make good use [3].

Vane Tip Overfiling and Underfiling

All impeller tip cuts and especially trim cuts across the entire impeller diameter (as done in the center illustration in Figure 11.3) produce blunt vane tips and hydraulic disturbances.

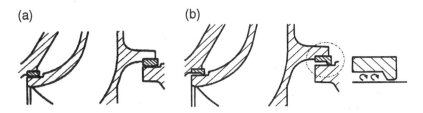

Figure 11.5 Conventional wear ring profiles (a) and enhanced contour wear rings (b). *Source:* Kurtz and Abraham [3].

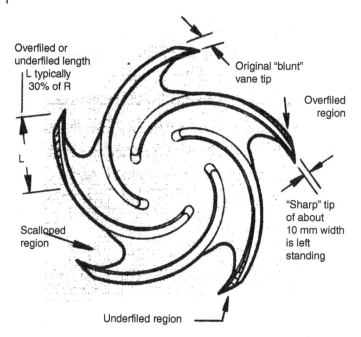

Overfiled or underfiled length
L typically 30% of R

Original "blunt" vane tip

Overfiled region

L

"Sharp" tip of about 10 mm width is left standing

Scalloped region

Underfiled region

Figure 11.6 Vane modifications suppress flow disturbances and vibrations.

These disturbances can be lessened by modifying the blunt vane tips by overfilling or underfiling (Figure 11.6).

Removing the metal from the leading edge of a vane tip is called "overfiling." The region from which metal is to be removed by tapering and blending-in is shown as "L". A guideline value for length "L" is 30% of the impeller radius R (or 15% of the impeller diameter). The vane tip width or thickness would be reduced to roughly 50% of its previous blunt edge width and metal removed so as to blend-in without creating a step or a ridge. So as not to compromise strength and resistance to erosion, even a "sharp" tip should not be thinner than 8 mm (~5/16th in.).

"Underfiling" is the term used for metal removal from the trailing edge of the impeller vanes. About 4% additional head can thus be gained near the BEP and the $H\text{--}Q$ curve is shifted slightly to the right. In other words, the liquid channel has been marginally enlarged and impeller performance is enhanced somewhat.

We chose a semiopen impeller (Figure 11.6) for ease of illustration only. It should be noted that scalloping – the removal of unsupported material between vanes – is generally done on the shroud (cover) of a closed impeller. This scalloping reduces resonant vibration and fatigue cracking risks [1].

Carbon Graphite Wear Rings and Bushings

Since about 2002, pump users have had incredible success upgrading critical process pump wear rings and bushings to Vespel® CR-6100. The proven track record of Vespel® CR-6100 has earned it a place in the American Petroleum Institute Standard API610, 11th edition, under its generic name perfluoro-alkoxy carbon-fiber reinforced composite, generally abbreviated as PFA/CF (see [4]). The recognized benefits of nonmetallic wear ring retrofits in pumps have users looking for other materials that will perform reliably in temperature applications greater than 230 °C (450 °F).

Carbon graphite is a proven candidate for wear rings and bushings in applications up to 1100 °F (A listing of the physical properties of carbon graphite can be found in Table 11.2). Carbon graphite is composed of a combination of amorphous carbon and graphite, which is obtained by baking a mixture of graphite with a carbon binder. During the baking process, outgassing occurs that results in porosity within the part, which is then filled with one of various available metals, depending on the physical properties required for the application (see [5]).

Table 11.2 Physical properties of carbon graphite and graphite material grades.

High temperature	Carbon graphite can withstand extremely high temperatures. Up to 1100 °F in oxidizing environments and 6000 °F in nonoxidizing environments.
Low temperature	Carbon graphite can withstand cryogenic temperatures without failing. At these temperatures, traditional lubricants and greases can congeal and solidify.
Dry environment	No additional lubrication is needed for carbon graphite parts in dry applications due to its self-lubricating properties.
Wet environment	Carbon graphite parts are self-lubricating and do not require additional lubrication in submerged applications where oil and grease lubricants would be washed away. Carbon/graphite will not swell as a result of the wet environment.
Self-lubricating	This is beneficial for high or low temperatures where oil and greases will not suffice. Eliminating the need for oil and grease also eliminates down time due to maintenance requirements related to oiling.
Dimensionally stable	Depending on processing, carbon/graphite parts can have an exceptionally low coefficient of thermal expansion; meaning that the size and shape of the part does not drastically change with a change in temperature. As a result, the material is able to withstand extreme temperatures (high and low) without deforming.
Corrosion resistant	Carbon graphite is chemically inert. As a result, carbon/graphite materials are able to perform well in corrosive environments where metal would corrode.

(Continued)

Table 11.2 (Continued)

Oxidation resistant	Oxidation is a common problem faced when carbon comes in contact with oxygen. Certain impregnations work to delay the onset of oxidation. This allows to extend the service life of the material in demanding conditions by increasing temperature resistance.
Wear resistant	Carbon graphite's self-lubricating properties coupled with improved thermal properties (low coefficient of friction and high thermal conductivity) results in better performance without excessive wear.
Low coefficient of friction	Coefficient of friction (COF) is related to heat generation. As objects slide against each other, the kinetic energy is converted to thermal energy. Carbon/graphite materials have a low COF when running against a counterface. As a result, less heat is generated during use.
High thermal conductivity	Thermal conductivity is a measure of a material's ability to conduct heat. Heat transfer occurs at a slower rate for materials with a low thermal conductivity. Graphite materials have a high thermal conductivity which allows heat that is generated due to friction to be transferred away from the contact point.
Food safe	There are a variety of impregnations that have been declared as Generally Recognized As Safe (GRAS) by the Food & Drug Administration (FDA), USA for use in sanitary or clean conditions such as the food, pharmaceutical, textile, paper, canning, and packaging industries. Some manufacturers can go a step further to obtain approvals from specialized agencies dealing with their products.
Electrically conductive	Depending on the impregnation, carbon graphite material grades can be electrically conductive, can eliminate static, will not spark, have very low electrical noise, disturbances in an electrical signal, allowing the material to transmit signals clearly.

Source: Metcar.com

A key benefit of using carbon graphite is its self-lubricating property, which allows it to be used in dry, high temperature, low temperatures, and food-safe applications without additional lubrication requirements. In addition, because carbon graphite parts are nonmetallic, they will not gall, which makes them ideal for use in low-viscosity liquid applications, like water, light hydrocarbons, and liquified gasses where hydrodynamic films are thin.

What We Have Learned

Doing things right would include allowing no shortcuts on impeller technology:

- Observing gaps "A" and "B" and doing oblique impeller trims is Best Practice. Inculcating a mindset that is uncompromising will pay back handsomely for decades. Best-in-class performers easily reach pump mean-times-between failure (MTBFs) of 6, 8, and even 10 years.

- Scalloping reduces the risk of crack formation and fatigue failure.
- Enhanced contour wear rings are simple and relatively effective performance enhancers.
- Some cooling fans are far too small to be of any value and cooling fans are often unnecessary on pumps.
- Making it a practice to allow deviations and accepting them will soon make deviations the "new norm." Allowing deviations to add up is a huge risk; it will certainly prevent a plant from ever becoming a best-in-class performer.

References

1 Dufour, John W., and William Ed Nelson; *"Centrifugal Pump Sourcebook"*, McGraw-Hill, New York, NY, 1993 (ISBN 0-07-018033-4).

2 Karassik, Igor J., William C. Krutzsch, Warren H. Fraser, and Joseph P. Messina; *"Pump Handbook"*, 2nd Edition, McGraw-Hill, New York, NY, 1985 (ISBN 0-07-033302-5).

3 Kurtz, Reinhard, and Katrine Abraham; "Special Systems for Coating Cars", 2. *World Pumps*, May 2010.

4 Bloch, Heinz; Fluid Machinery: Life Extension of Pumps, Gas Compressors and Drivers, Chapter 8-Why consider perfluoro-alkoxy carbon-filled polymers, Published by De Gruyter, 2020.

5 Hardyal, K.; *"Carbon Graphite Technical Bulletin"*, Metallized Carbon Corporation, 2021.

12

Lubrication Management

A listing of the four or five worst enemies of proper and reliable lubrication must include lubricant contamination, viscosity, and application deficiencies. These three parameters are interrelated. For instance, water entering as an outright contaminant or in the form of air saturated with moisture often causes additive depletion. The lubricant then rapidly degrades and bearings fail prematurely. Water can also act as the catalyst for sludge formation and often has an undesired effect on viscosity. Oil rings are usually designed for ISO VG 32 oils tend to malfunction in oils of inappropriate viscosity.

Profitability-minded plants pay attention to these facts and thoughtfully manage all of their lubrication practices, including oil storage and oil transfer. With thoughtful lube management, they reap substantial benefits from the enhanced pump reliability that results.

How Bad Is Water Contamination?

Even relatively small amounts of water contamination tend to affect bearing performance. The life of an antifriction bearing with even as little as 100 ppm of free water in the lubricant (1 teaspoon of water in 13 gal of oil) will be only 40% of the life of the same bearing with 25 ppm free water present [1]. Also, most major bearing manufacturers have published estimates on the effects of lubricant contamination [2].

Considerable time is often wasted in arguments over how much water should be tolerated in pump bearing housings. It would be far more productive to locate the source of water intrusion and to cut off the source. Certainly, the amount of water drained from the pump bearing housing in Figure 12.1 is totally unacceptable.

Pump Wisdom: Essential Centrifugal Pump Knowledge for Operators and Specialists,
Second Edition. Robert X. Perez and Heinz P. Bloch.
© 2022 The American Institute of Chemical Engineers, Inc. Published 2022 by John Wiley & Sons, Inc.

Figure 12.1 Excessive amount of water being drained from a pump bearing housing.

Figure 12.2 Unsuitable lube transfer containers.

In all instances, getting rid of unsuitable transfer containers (Figure 12.2) must be made one of the first orders of business.

Galvanized oil transfer containers are frequently attacked by certain lube oil additives. Good lube management practices mandate the use of approved

Figure 12.3 Careless handling of an approved transfer container can still cause extreme particulate contamination. *Source:* Fluid Defense Systems [3].

dispensing containers. Of course, even the best purpose-designed plastic containers can be defeated by careless handling. This is illustrated in Figure 12.3.

Lube oil contamination can be kept in check in several ways. Modern bearing protector seals will prevent the ingress of water, steam, and other airborne contaminants (see Chapter 8). Either dual face (contacting) magnetic seals or advanced rotating labyrinth style (noncontacting) bearing protector seals are available. Both types make much economic sense and merit consideration in new and existing (retrofit-to-upgrade) installations.

Avoid Solids Contamination

In Figure 12.4 and similar illustrations published in technical texts, major multinational bearing manufacturers alert us to the life reduction risk associated with lube oil contamination. The ratio of operating viscosity of the lubricant (see Figure 7.2) over its rated lubricant viscosity (see Figure 7.1) is calculated as v/v_1 and entered on the x-axis.

Three relative stages of oil cleanliness are indicated by three different regions. Region I is applicable only to situations combining utmost cleanliness in the lubrication gap with moderate bearing loading. This situation is rather unrealistic for process pumps. Region II is where a high degree of cleanliness is maintained; this condition is assumed achievable with modern bearing protector seals (see Chapter 8). Finally, assume Region III might represent bearing housing interiors exposed to an industrial environment or bearing housings *without* effective

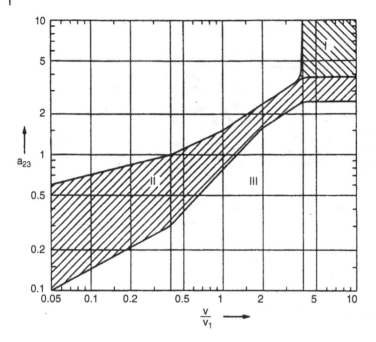

Figure 12.4 Contamination adjustment factor "a_{23}" vs. viscosity ratio v/v_1.
Source: Modified from SKF USA, Inc. [2].

protector seals. A resulting life adjustment multiplier or factor "a_{23}" is displayed on the y-axis.

Using Figure 12.4 and for, say, $v/v_1 = 0.5$, one would obtain $a_{23} = 0.3$ in Region III, and $a_{23} = 1$ in Region II. In essence, using effective bearing protector seals (see Chapter 8) yields an anticipated threefold bearing life improvement.

Avoid Questionable Oil Storage and Transfer Practices

Questionable drum storage practices are shown in Figures 12.5 and 12.6. Changes in ambient temperature cause rainwater to be drawn into the drum by capillary action. Capillary action takes place even in the case of seemingly tight bungs (Figure 12.5) as long as rainwater is allowed to collect in the drum top. The drum top will bulge outward (convex shape) when the surrounding air warms up; it will bulge down (concave shape) when the air cools and a slight vacuum is produced in the vapor space inside the drum. The resulting suction effect draws in water.

Open and up-turned vents in Figure 12.6 invite entry of both moisture and particulate contamination. Each drum contains US$1200 worth of synthetic

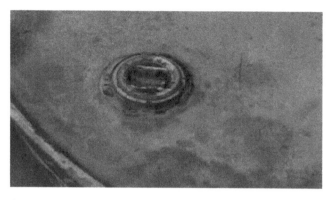

Figure 12.5 Rainwater will enter through this closed bung via capillary action.

Figure 12.6 Unacceptable outdoor lube storage where up-turned open vents allow contaminants and moisture to enter.

lubricant that is now rendered unserviceable. One solution is shown in Figure 12.7 where outdoor drum storage has been replaced by indoor mini-bulk storage in sheltered or similarly protected locations. Bulk tanks and drums are placed under cover.

Contamination control of bulk storage containers makes it highly desirable to put oil drums indoors. In the event that outdoor storage is chosen, the storage area must be covered by a suitable roof and side panels to ward off direct rain and snow.

Largely self-contained lubrication work centers (Figure 12.8) are available for clean oil storage and oil dispensing. These also merit serious consideration [3].

Figure 12.7 Mini bulk storage indoors. Note approved transfer containers on upper shelf of cabinet. *Source:* Fluid Defense Systems [3].

Figure 12.8 Modern lubrication storage center, about 2010.

They replace traditional storage in cumbersome and generally inefficient 55 gal drums. Modern lubrication work centers incorporate an interchangeable series of frames, pumps, filters, and storage modules. There also are spill containment cans and other suitable keep-clean provisions.

The new work centers should be preferred over antiquated tank and drum rack systems. Older systems often pay little attention to contamination avoidance and a facility's lubrication work flow processes.

Periodic Audits

Periodic audits of lubrication practices are performed at leading-edge facilities and many of these lube audits uncover unsatisfactory lubricant dispensing practices. Audit findings and recommendations often involve detection of oil contamination in sump-lubricated equipment (Figure 12.1) and the labeling of points to be lubricated.

Regrettably, some plants persist in arbitrarily "standardizing" on less-than-optimum oil and grease formulations. Many are employing incorrect regreasing practices on the millions of electric motors used to drive process pumps. Superior plants experience as few as 14 motor bearing replacements per 1000 motors per year; average plants replace 156 motor bearing replacements per 1000 motors per year [1]. The statistics of less-than-average plants are far worse.

Suffice it to say that being unaware of best available lubrication management and practices can be very expensive. These expenditures could be avoided by simply keeping out contaminants. Experience shows that periodic lube management audits are cost-effective and almost always point out areas of improvement.

What We Have Learned

- Correct outdoor storage of oil drums involves considerable forethought and constant attention.
- Storage under a shelter or indoor storage is much preferred over full exposure to the ambient environment.
- Water and particulate contaminants can greatly impair bearing life. They must be kept away from pump bearings.
- Drum vent provisions are needed, but open vents that allow water and dirt ingress are unacceptable.
- Transfer containers should be purposefully designed and be kept immaculately clean.
- Periodic lube management audits are recommended.

References

1 Bloch, Heinz P.; *"Machinery Reliability Improvement"*, 3rd Edition, Gulf Publishing Company, Houston, Texas, 1998 (ISBN 0-88415-661-3).
2 SKF USA, Inc.; Kulpsville, Pennsylvania, General Catalog 4000 US, 1991.
3 Fluid Defense Systems; Montgomery, Illinois 60538, 2010.

13

Pump Condition Monitoring

Pump Vibration, Rotor Balance, and Effect on Bearing Life

Process pump user-operators often want simple rules-of-thumb to determine maximum allowable vibration. Of course, rules-of-thumb should not be confused with statistical proof. Many times, general experience and common sense are of greater value than statistics. Essentially, this chapter deals with a few experience-based observations on the issue of condition monitoring for process pumps [1].

Vibration and Its Effect on Bearing Life

Pumps, like all rotating machines, vibrate to some extent due to response from excitation forces, such as residual rotor unbalance, turbulent liquid flow, pressure pulsations, cavitation, and pump wear. The magnitude of vibration will also be amplified as flows deviate from best efficiency (Figure 13.1) and as the vibration frequency approaches the resonant frequency of a major pump, foundation, and/ or connected piping.

Vibration from the running pump is very often transmitted to the nonrunning (standby) pump. This transmitted vibration tends to wipe off the oil film on the bearings of nonrunning pumps, causing metal-to-metal contact. Bearing degradation then shows up when the standby pump is put in service. The rate of degradation is often reduced by switching or alternating from the "A" pump to the "B" pump on a four- to six-week basis.

A number of published observations on pump vibration and its effect on bearing life lead to a plot (Figure 13.2) which probably brackets 90% of all process pumps.

No two predictions are the same; yet, Figures 13.2 and 13.3 illustrate the same point: Vibration excursions tell a story and reflect the condition of a process pump. The root cause may be hydraulic and temporary; it could relate to flow

Pump Wisdom: Essential Centrifugal Pump Knowledge for Operators and Specialists, Second Edition. Robert X. Perez and Heinz P. Bloch.
© 2022 The American Institute of Chemical Engineers, Inc. Published 2022 by John Wiley & Sons, Inc.

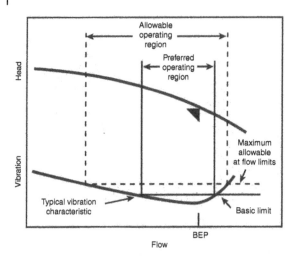

Figure 13.1 Prominent pump standards recognize pump vibration magnitude is not uniform over the entire flow range. *Source:* API-610, 8th Edition, 1996.

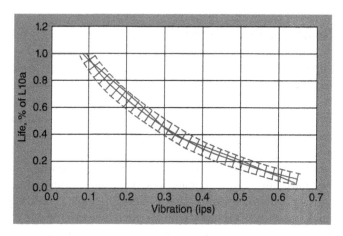

Figure 13.2 How pump vibration affects bearing life. *Source:* Bloch and Budris [2].

disturbances that vary with flow rate, or the root cause could be related to deficiencies in one or more mechanical components. Component imbalance or bearing defects may be causing vibrations. Either way, vibration reduces bearing life in accordance with Figure 13.2.

The absolute value of vibration is not necessarily as important as the suddenness of a vibration increase. As an example, if pump vibration had been around 0.1 ips (2.5 mm/s) for the past two years and had increased to 0.3 ips (7.5 mm/s) in a single day, we could consider this a more serious event than vibration that

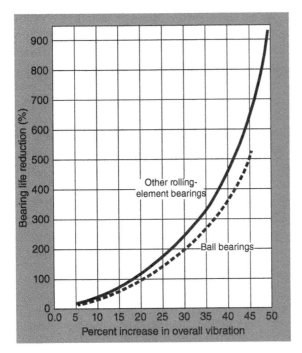

Figure 13.3 Bearing degradation shows up as vibration and can be related to life reduction.

started at 0.3 ips and then gradually increased to 0.4 ips (10 mm/s) in the span of 12 months.

Figure 13.3 relates (conservatively) how an increase in overall vibration due to bearing-internal deterioration will shorten bearing life and by what percentage. It can be reasoned that vibratory activity adds to the normal bearing load. Bearing manufacturers report that rolling element bearing life varies exponentially with load. For a typical ball bearing, the exponent is 3; therefore, a twofold load reduces bearing life by a factor of $2^3 = 8$. The bearing life is then only 12% of what it would have been at normal load conditions. A compelling case is thus made for keeping vibration low.

For most pump reliability improvement professionals, the issues of interest are not whether or not the pump vibrates but

- If the amplitude and/or frequency of the vibration is sufficient to cause actual or perceived damage to any of the pump components
- If the vibration is a symptom of some other damaging phenomenon happening within the pump
- If the relationship between vibration severity and bearing life can be quantified with a reasonable degree of accuracy

Various industry organizations, such as the Hydraulic Institute (HI, in ANSI/HI 9.6.4, [3]), and the American Petroleum Institute (API in its Standard API-610) have set pump vibration limits for general guidance. All are reaching back to the experience of individual reliability professionals and multinational pump user companies that had implemented daily machinery condition reviews (monitoring and surveillance) decades ago.

Some of these professionals had assisted operating personnel by listing acceptable, reportable, and mandatory shut-down levels of vibration. These levels represent experience-based values that rely on the bearing life vs. vibration approximations for general-purpose machinery. They have been widely published since the late 1940s as allowable vibration velocities.

Traditionally, 0.35 ips rms (~9 mm/(s rms)) was given as a maximum allowable vibration velocity for "total all pass" (overall) readings taken on bearing caps or housings. However, machinery vibration and its measurement are complex matters and may require some clarification. Typical considerations might include the following:

1) Vibration can be measured and/or analyzed by using units of displacement, velocity, or acceleration severity to evaluate the health of the machine. As stated earlier, the primary or traditional measure of vibration used by industry today is velocity. Because most pump users use this parameter, comparisons are made easier.

2) Either the "total all-pass" or the "filtered" frequency can be used. Most industry specifications and standards use "total all-pass" vibration values to identify problem pumps. Filtered values are reserved for determining where the vibration originates. This latter determination is generally called "vibration analysis."

3) Root mean square (RMS) as well as peak-to-peak values are sometimes measured or specified. The Hydraulic Institute (HI) has chosen RMS acceptance limit values. HI recognizes that most vibration instruments actually measure vibration in RMS terms and then calculate peak-to-peak values, if required. API, on the other hand, generally refers to peak-to-peak readings. RMS values are roughly 0.7 multiplied by peak-measured values.

   However, this relationship applies only to vibration consisting of a single sinusoidal waveform. For more complex waveforms, this conversion does not yield correct results.

4) The acceptable vibration amplitude (as-new vs. postrepair levels) may have to be specified for a particular application. Acceptance limits will change along with overall pump power and flow rate regions. The Hydraulic Institute and ISO base acceptable vibration limits on pump type and power level, while API gives different limits for the "preferred" and "allowable" operating regions. (See vibration acceptance limits, below, for "as-new" acceptance values).

5) It is difficult to predefine how factory test stand vibration measurements should compare with field (at site) values. The exact effects of foundation stiffness/grouting are difficult to predict. Generally, the stiffness of a field pump foundation is much higher than the stiffness found on a factory test stand, especially if the pump base plate is grouted. That is why the Hydraulic Institute vibration standard allows higher test stand values (up to twice field values).

   For vertical turbine pump installations, it is especially important to know the actual foundation stiffness to avoid high vibration from operation at a structural resonance frequency.

6) Cataloging how much the vibration amplitude and/or frequency has changed over the life of the machine is important. This is called "trending." It is especially helpful to have an as-new vibration signature taken and kept on file for future comparison.

7) Location of vibration measurements: On a typical horizontal process pump, vibration readings are taken in the x, y, and z (axial) directions.

   Horizontal and vertical dry pit pump vibration measurements are normally taken on or near the outer, or uppermost, bearing in the horizontal, vertical, and axial planes, with the maximum value used for acceptance.

   Vertical turbine pump vibration measurements are taken at the top or bottom of the motor. Probes should not be located on flexible panels, walls, or motor end covers.

Monitoring Methods Differ

Some data gathering methods employ shock pulse monitoring (SPM) and, like all of the other methods, may or may not forward the results by wireless means. In simple terms, the SPM method detects the development of a mechanical shock wave caused by the impact of two masses. At the exact instant of impact, molecular contact occurs and a compression (shock) wave develops in each mass. The SPM method is based on the events occurring in the mass during the extremely short time period after the first particles of the colliding bodies come in contact. This time period is so short that no detectable deformation of the material has yet occurred. The molecular contact produces vastly increased particle acceleration at the impact point. The severity of these impacts can be plotted, trended, and displayed.

There is also temperature monitoring. Suppose a pump bearing housing operates at 170 °F (~77 °C). Most people can place an index finger on such a bearing housing for about five seconds before the pain becomes too intense. However, 170 °F (~77 °C) is not excessive for process pumps (see Chapter 8). Proper pump surveillance calls for measurements with either a surface pyrometer or a handheld noncontacting infrared heat-sensing instrument (thermal gun).

Under no circumstances should the bearing housing be doused with water. Such cooling would probably cause the bearing outer rings to be cooled. Because metal shrinks upon being cooled, the already small bearing-internal clearances would be reduced to the point of being excessively preloaded. Bearing life would be curtailed by the very method thought to extend it.

Vibration Acceptance Limits

Hydraulic Institute Standard ANSI/HI 9.6.4 presents the generally accepted allowable pump "field" vibration values for various pump types (see Table 13.1). The standard is based on RMS total, or all-pass vibration values. The standard states that factory or laboratory values can be as much as twice these field limits, depending on the rigidity of the test stand. The ANSI/ASME B73 standard accepts two

Table 13.1 Allowable field-installed vibration values for pumps.

Pump type	Less than power (HP)	Vibration RMS (in./s)	Greater than power (HP)	Vibration RMS (in./s)
End suction ANSIB73	20	0.125	100	0.18
Vertical inline, separately coupled, per ANSI B73.2	20	0.125	100	0.18
End suction and vertical inline close-coupled	20	0.14	100	0.21
End suction, frame-mounted	20	0.14	100	0.21
End suction, API-610, preferred operation region (POR)	All	0.12	All	0.12
End suction, API-610, allowable operation region (AOR)	All	0.16	All	0.16
End suction, paper stock	10	0.14	200	0.21
End suction solids handling – Horizontal	10	0.22	400	0.31
End suction solids handling – vertical	10	0.26	400	0.34
End suction hard metal/rubber-lined, horizontal and vertical	10	0.30	100	0.40
Between bearings, single and multistage	20	0.12	200	0.22
Vertical turbine pump (VTP)	100	0.24	1000	0.28
VTP, mixed flow, propeller, short set	100	0.2	3000	0.28

Source: Modified from ANSI/HI Standard 9.6.4 [3].

times the HI9.6.4 values for factory tests performed on chemical end suction pumps [3]. HI includes the API-610 values for end suction refinery pumps (in RMS terms); the API-610 document requires that these acceptance values be demonstrated on the factory test stand.

The HI Standard also states that stipulated values only apply to pumps operating under good field conditions. Good field conditions are defined as follows:

1) Adequate NPSH (net positive suction head) margin.
2) Operation within the pump's preferred operating region – typically 70% and 120% of BEP (Table 13.3 only lists the constant values required for low- and high-pump power ratings. The acceptable vibration, between the low- and high-power values, varies linearly with power on a semilog graph).
3) Proper pump/driver shaft (coupling) alignment.
4) Pump intake must conform to ANSI/HI9.8 ("Pump Intake Design").

It should also be noted that the acceptable vibration values for slurry and vertical turbine pumps are about double the values given for horizontal clean liquid pumps.

Once a pump is accepted and commissioned, somewhat higher total (all-pass) vibration values are usually accepted before further follow-up and analysis are deemed appropriate. As a general rule, repair follow-up is recommended if vibration levels increase to twice the "field" acceptance limits (or initial actual readings).

Causes of Excessive Vibration

Once a pump has been determined to have a high "total or all-pass" vibration level, the next step is to identify the cause. This would be the time to obtain a filtered vibration analysis and to look for predominance of one of many frequencies in the spectrum. Table 13.2 illustrates several predominant frequencies, although it is providing a narrow overview, at best.

Along those lines, the first step in the analysis should be to capture, and then evaluate, the multiples of pump running speed (Table 13.2). A graphic display

Table 13.2 Sources of specific vibration excitations.

Frequency	Source
$0.1 \times$ Running speed	Diffuser stall
$0.8 \times$ Running speed	Impeller stall (recirculation)
$1 \times$ Running speed	Unbalance or bent shaft
$2 \times$ Running speed	Misalignment
Number of vanes $\times$ running speed	Vane/volute gap and cavitation

would often be called a "filtered" velocity plot, or frequency spectrum. Actual analysis can point to several possible causes, among them:

1) Rotor unbalance (new residual impeller/rotor unbalance or unbalance caused by impeller metal removal – wear).
2) Shaft (coupling) misalignment.
3) Liquid turbulence due to operation too far away from the pump best efficiency flow rate
4) Cavitation due to insufficient NPSH margin.
5) Pressure pulsations from impeller vane – casing tongue (cutwater) interaction in high discharge energy pumps.

Other possible causes of vibration may be more complex to analyze. Among these are the following: Operating speed close to mechanical or hydraulic resonant frequencies of a major pump, foundation, or pipe component. This is of special importance with large multistage horizontal and long vertical pumps. A margin of safety should be provided between rotor and/or structural natural frequencies and operating speed. Typical margins are 15–25%. Vibration amplification will generally be greater than 2.5-times at a resonant frequency.

Vibration/resonance events to be evaluated on pumps include rotor lateral vibration and structural lateral vibration – rather common with long-shafted vertical pumps.

Poor pump suction or discharge piping can also cause increased vibration, normally by either increased cavitation or turbulent flow in the pump. Pump operating speed or vane pass frequencies could excite a piping structural or hydraulic resonance. (For additional comments on piping vibration see Chapter 4).

Bearing wear will usually show up in the vibration spectrum. Rolling element bearings have distinct vibration signatures based on the number of bearing balls or rollers. Recall, however, that monitoring deterioration of plastic bearing cages would require highly sophisticated monitoring techniques. This is one of the reasons why plastic cages can be used in pump bearings only after all relevant factors are taken into account.

Opening up of impeller wear ring clearances is primarily shown in performance measurements [1]. This wear can reduce the $NPSH_r$ margin and shift the pump operating flow point.

Broken rotor bars in electric motors will generate specific frequencies.

Rotor Balancing

All impellers, irrespective of their operational speed, should be dynamically balanced ("spin-balanced") before installation, either single or two plane. Two-plane balance is required for a wide impeller, typically when the impeller width is greater than 17% of the impeller diameter. ISO balance criteria are usually invoked and an experienced balance shop will know them well.

Dynamic balancing of the three major rotating pump components, shaft, impeller, and coupling, will increase mechanical seal and bearing life. All couplings in the weight or size ranges found in a modern refinery should be balanced, if they are part of a conscientious and reliability-focused pump failure reduction program.

Of course, if a facility is willing to remain repair-focused, it can continue to just plod along with "business as usual." Still, reliability-focused plants agree: large couplings that cannot be balanced have no place in the majority of their process pumps.

The preferred procedure for process pumps in reliability-focused installations is to balance the impeller and coupling independently, and to then balance the impeller and coupling on the shaft as a single unit. Another method is to balance the entire pump rotor as an assembled unit and to do so one time only. That might be a bit problematic at locations that will subsequently go through repair cycles while trying to omit full rotor balance. Often, more problems are caused in successively disassembling and reassembling than would be caused by diligently balancing each individual component. For multistage pump rotors (both horizontal and vertical), individual component balance is generally preferred.

The static (single plane) balance force is always the more important of the two forces, static and dynamic ("couple force"). If balancing of individual rotor components is chosen, it is best to use a tighter tolerance for the static (single plane) force. In theory, if all the static force is removed from each part, there should be very little dynamic (couple) force remaining in the rotor itself.

For impellers operating at 1800 rpm or less, the ISO 1940 G6.3 tolerance is acceptable. For 3600 to 1800 rpm, the ISO G2.5 rule is better. Both are displayed on balance tolerance nomograms for small and large machinery rotors. Generally, tighter balance tolerances (G1.0) are not warranted unless the balancing facility has modern, automated balancing equipment that will achieve these results without adding much time and effort.

Using older balancing equipment may make it difficult and unnecessarily costly to obtain and duplicate the G1.0 quality. Also, factory vibration tests have, at best, shown insignificant reductions in pump vibration with this tighter balance grade.

That, however, is not the point. Instead, let us realize that relatively tight balance tolerances or good grades of balance are obtained on automated balancing machines just as quickly as would more liberal, less precise, balance specifications. Using an analogy, why allow bottles of medicine to contain between 99 and 101 tablets when modern filling machinery can guarantee to deliver precisely 100 tablets per bottle? Surely, a serious and reliability-focused user-consumer will insist on products with consistently high quality.

Balancing machine sensitivity must be adequate for the part to be balanced. This means that the machine should be capable of measuring unbalance levels to

Table 13.3 Maximum diametral hub looseness.

Impeller hub bore	Maximum diametral looseness	
	<1800 rpm	1800–3600 rpm
0–1.499 in.	0.0015 in.	0.0015 in.
1.5–1.999 in.	0.0020 in.	0.0015 in.
2.0 in. and larger	0.0025 in.	0.0015 in.

one-tenth of the maximum residual unbalance allowed by the balance quality grade selected for the component being balanced.

Rotating assembly balance is recommended whenever practical and if the tighter quality grades, G2.5 or G1.0, are desired. Special care must be taken to ensure that keys and keyways in balancing arbors are dimensionally identical to those in the assembled rotor. Impellers must have an interference fit with the shaft when G1.0 balance is desired. Although looseness between impeller hub and shaft (or balance machine arbor) is allowed for the lesser balance grades, it should not exceed the values given in Table 13.3 for grades G2.5 or G6.3.

What We Have Learned

- Initial guidance on allowable pump vibration is clearly available from the hundreds of articles and dozens of books that have been published in the decades since 1960. Up-to-date summaries are contained in [2] and other modern texts.
- Since elevated vibration increases the forces acting on bearings, and since bearing life is related to bearing load, higher vibration will reduce bearing life. The rules of thumb and empirical relationships express these guidelines with sufficient accuracy for general-purpose equipment.
- Modern data collectors and condition analyzers are available from a number of competent vendor-manufacturers. Also, many models are operating with wireless connections, while others are handheld or hard-wired [4]. Each has its advantages, and these must be considered on a case-by-case basis.
- Since rotor unbalance will lead to increased vibration, good rotor balance is essential. The issue of balance grade is moot in the very many instances where modern, often fully automated balancing machines are readily available. These balancing machines will achieve excellent equipment rotor balance as quickly and effectively as not-so-excellent balance.
- Bearing life is related to shaft misalignment and force transmission across couplings. These affect vibration severity [5]. While rules of thumb are not absolutes, their judicious application makes far more sense than rather simplistic requests to "prove it to me."

The issue is about risk and the mitigation of risk. A fitting analogy deals with automobiles, where reasonable people know that driving on worn tires will put the passengers at greater risk than driving on new tires.

Likewise, the issues of vibration and shaft misalignment are intuitively evident to most of us. We should be quite satisfied with rules of thumb and empirical data where they appeal to common sense. They certainly do in this instance.

Don't compromise safety and reliability. Keep process pump vibration low.

References

1 Beebe, R.S.; "Machine Condition Monitoring", MCM Consultants, Monash University, Gippsland, Australia, Engineering Handbook (2001 Edition).
2 Bloch, Heinz P., and Alan Budris; "*Pump User's Handbook*", 3rd Edition, Fairmont Press, Lilburn, GA 30047, 2010 (ISBN 0-88173-627-9).
3 ANSI/HI Standard 9.6.4; Hydraulic Institute, Parsippany, NJ, 2001.
4 Bloch, Heinz; Paul Lahr; Donald Hyatt; "Development of an Advanced Electronically Optimized Variable High-Speed Centrifugal Pump", Proceedings of Pump Congress, Karlsruhe, Germany, October 4–6, 1988.
5 Berry, Douglas L.; "Vibration vs. Bearing Life Increase", *Reliability*, December 1995.

14

Drivers, Couplings, and Alignment

Recall, please, that this text deals with pump-related issues that are often over-looked. Drivers, couplings, and the accuracy of shaft alignment are pump-related and affect availability and operating life. They deserve our attention.

Driver Selection

Driver selection is based on criteria that involve many different considerations, including commercial and in-plant power grids, substations, etc. Types of winding insulation are specified by the design contractor, but the power demand of a particular process pump goes up during episodes of pump overload or as pump internals wear. These load increases are not always addressed in the motor sizes being ordered.

For a pump requiring 125 hp input, buying a 150 hp motor with a service factor (S.F.) of 1.0 will provide a longer ultimate motor life than buying a 120 hp motor with an S.F. of 1.15. Proper motor selection should also take into account temperature profiles and installation geography, i.e. elevation above sea level. All can influence motor efficiency and winding life.

Motor lubrication should either be specified by the purchaser or, alternatively, be discussed with the motor vendor. This discussion and approval step is especially important with oil-mist lubrication, where a special type of insulating tape must be used in the junction box and on T-leads [1]. Other than that, oil-mist lubrication on motors has been highly successful since the mid-1960s and, on rolling element bearings, is much preferred over other lubrication methods. The ingress of pure oil mist into a modern electric motor will not adversely affect it. The slightly pressurized mist environment may, in fact, keep out airborne dirt.

Pump Wisdom: Essential Centrifugal Pump Knowledge for Operators and Specialists,
Second Edition. Robert X. Perez and Heinz P. Bloch.
© 2022 The American Institute of Chemical Engineers, Inc. Published 2022 by John Wiley & Sons, Inc.

Approval and disclosure of the path traveled by grease through a motor bearing is also important. Many times, electric motor bearings are not arranged for effective relubrication. The intent of shielded bearings (whereby the replenishing grease actually enters the rolling elements through capillary action and *not* through direct pressurization!) is sometimes poorly understood by all parties. Application of undue pressure to a bearing shield can cause the rolling elements to scrape on the shield. Injecting grease at excessive pressure very often hastens motor bearing failures [2].

Coupling Selection and Installation

The lowest initial cost coupling is rarely the best choice once life cycle cost and safety risk are put into the equation. Process pump couplings should have a service factor (S.F.) of 2 or 2.5. Elastomeric couplings designed with the flexible element twisting or pulling should not be used for large pumps. A number of factors come into play here:

- Toroidal ("tire-type") flexing elements will exert an axial pulling force on driving and driven bearings. Also, they are difficult or impossible to balance.
- Polyurethane flexing elements perform poorly in concentrated acid, benzol, toluene, steam, and certain other environments.
- Polyisoprene flexing elements do poorly in gasoline, hydraulic fluids, sunlight (aging), silicate, and certain other environments.

Gear couplings require grease replenishment and are certainly more maintenance-intensive than nonlubricated disc pack couplings. If gear couplings are selected, they should be on the plant's preventive maintenance schedule. Only approved coupling greases should be used for periodic replenishment. Realize that standard and multipurpose greases are unsuitable because their oil and soap constituents get "centrifuged apart" at typical pump coupling peripheral speeds.

That leaves a reliability-focused user with nonlubricated disc pack (Figure 14.1) or other variants of alloy steel membrane couplings as the preferred choice. The design must be such that the spacer piece or center member is captured. In the unlikely event of a disc pack failure, a "captured center member" will not leave the space between the two coupling hubs. Many brands of cheap couplings do not have captured center members.

Installation and Removal

Before installing a coupling, examine it for adequacy of puller holes or other means of future hub removal. The coupling in Figure 14.2 was mistreated at disassembly because no thought had been given to future removal.

Figure 14.1 A typical captured center member disc pack coupling for process pumps.

Figure 14.2 Coupling hub damaged with severe hammer blows at disassembly.

For parallel pump shafts with keyways, use 0.0–0.0005″ (0.0–0.012 mm) total shaft interference. Use one of several available thermal dilation methods (heat treatment oven, superheated steam, electric induction heater) to mount hubs on shafts.

Disallow loose-fitting keys for coupling hubs because they tend to cause fretting damage at shaft surfaces. During rebuilding or repair, allocate time needed for hand-fitting keys; they should fit snugly in keyway. On all replacement shafts, machine radiused keyways and modify the keys to match the radius contour.

Alignment and Quality Criteria

Pump baseplates should incorporate means of adjusting pumps and drivers for proper shaft alignment. Jacking bolts for horizontal movement in two directions were discussed in Chapter 3 and are also shown in Figure 14.3. Shaft alignment

Figure 14.3 A quality process pump during factory assembly. Note alignment jacking bolts on pump supports and on motor. *Source:* Emile Egger & Cie., Cressier, NE, Switzerland.

work can be carried out with the bearing housing support brackets bolted up securely in this instance. Here, the pump's two large size main support pedestals and its bearing support bracket can reasonably be expected to operate at the same temperature.

But suppose a pump casing had mounting feet integrally cast with the pump casing (Figure 14.4), and these integral mounting feet would be bolted to the base plate. In that case, the thermal rise of the (hot) pump might differ from that of the (somewhat colder) ambient air-flooded bearing support bracket. The bracket would have to be unbolted while shaft alignment work is in progress. The bracket needs to be reconnected *after* the pump has been started and has reached temperature equilibrium.

Note that the pump model shown in Figure 14.5 does not incorporate a bearing support bracket. Designing a sturdy pump without the need for a bearing support bracket circumvents the issue mentioned in the previous paragraphs. (Do not overlook that a coupling guard must be installed before allowing this pump – or any other pump – to be put into service. An epoxy prefilled base plate with motor jacking provisions is shown on this fully mounted pump set).

A number of different alignment methods exist; the reverse indicator method setup of Figure 14.6 is still widely used. Reverse dial indicator alignment is sound

Bearing support
bracket

Figure 14.4 Foot-mounted ANSI pump on a conventional base plate. On other than ambient temperature services the bearing support bracket must be left unbolted, while the pump grows up or down to reach thermal equilibrium at its operating temperature.

Figure 14.5 This medium-size pump is designed without a bearing support bracket. It incorporates motor jacking bolts and is shipped fully aligned on an epoxy-filled base plate [3]. *Source:* Fairmont Press Inc.

and appropriate. However, it must be properly executed and due attention given to indicator bracket sag. As is the case with all pump alignment methods, the task encompasses cold-aligning with offsets to compensate for equipment thermal growth. Alignment work crews must be well-trained and conscientious.

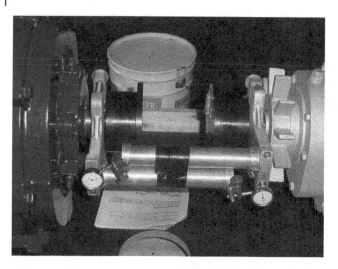

Figure 14.6 Reverse dial indicator alignment setup using low-sag tubular indicator mounting brackets [3]. *Source:* Fairmont Press Inc.

The reverse dial indicator method forms the basis of its successor technology: modern laser alignment techniques. Because laser alignment is both fast and precise, it has become the preferred method. It is used extensively by Best Practices Companies.

Consequences of Misalignment

Pump misalignment has serious consequences. Among other shortcomings misalignment may be forcing standard bearings to operate at an unintended angle. The simplest approximation of misalignment is obtained by measuring the parallel offset between a pump shaft and the shaft of its driver. Also, the distance between the two shaft ends is measured. The tangent is a trigonometric function; it is obtained by dividing the shaft offset (inches or mm) by the distance between shaft ends (inches or mm), often labeled DBSE (distance between shaft ends).

To achieve the rated bearing life of 100%, [4] recommends keeping the tangent of the misalignment angle at 0.001 and lower. This reference describes tests carried out on a variety of bearings, and for most of these, the effect of misalignment on life is described by the boundaries of the two curves in Figure 14.7. This graphic describes that once the tangent exceeds 0.006, rolling element bearing service life drops into the 10–20% range. However, aiming for a tangent value of 0.001 is

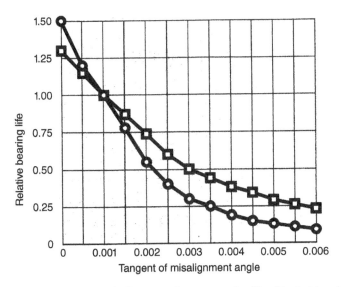

Figure 14.7 How misalignment shortens service life of typical bearings. Most bearings fit somewhere between the two curve boundaries. At tangents below 0.001, bearing life is here assumed to exceed a relative rating of 1. *Source:* Leibensperger [4].

rather generous compared to the considerably more stringent tangents found at some Best-of-Class pump user companies. Best-of-Class practice is not to allow shaft misalignments in excess of 0.5 mils per inch (0.0005 mm/mm) of shaft separation. This rather severe guideline makes allowance for angular misalignment that may exist in addition to parallel offset.

Thermal Rise and Predefinition of Growth

We assume that process pumps are driven by electric motors or small steam turbines. As these drivers startup, they also warm up. Of course, they "grow" as the temperature increases, and this thermal growth must be taken into account when aligning driver and driven shafts.

Because a typical electric motor driver does not grow too hot, its thermal growth can be neglected. The same is true for foot-mounted process pumps with product temperatures up to about 200 °F (~93 °C). On foot-mounted pumps with process temperatures in excess of 200°°F, the cold offset (in inches) can be calculated from

$$\Delta H = \left(0.000006\right)\left(H\right)\left(\Delta T\right)$$

In this formula, ΔH is the height difference below which the pump shaft should be set in the cold condition. H is the height or distance from the bottom of the feet of the pump to its centerline and ΔT is the temperature (°F) of the pumped fluid after subtracting 200.

Here is an example for a pump with a distance from the bottom of the feet to the shaft centerline equal to 18 in. and a process temperature of 430 °F:

$$\Delta H = (0.000006)(18)(430 - 200) = 0.025 \text{ in.}$$

This pump is expected to grow about 0.025 in. due to being heated 230 °F hotter than the driver. Of course, this is only an approximation, but it will work.

On API-style centerline-mounted pumps (Figure 13.5) with process temperatures in excess of 200 °F, the offset can be calculated from

$$\Delta H = (0.000002)(H)(\Delta T)$$

In this formula, ΔH is the height difference below which the pump shaft should be set in the cold condition. H is the height of the pedestal (usually the distance from the base plate to the pump centerline), and ΔT is the temperature of the pumped fluid after subtracting 200.

Here is an example for a pump with a pedestal height of 21 in. and a process temperature of 740 °F:

$$\Delta H = (0.000002)(21)(740 - 200) = 0.022 \text{ in.}$$

This pump is expected to grow about 0.022 in. due to being heated 540 °F hotter than the driver. This, too, is only an approximation, but it will work.

Aim not to allow shaft misalignment to exceed the limits plotted by competent alignment service providers. Use 0.5 mils per inch (0.0005 mm/mm) of shaft separation (DBSE, or distance between shaft ends) as the maximum allowable shaft centerline offset.

In the many decades since 1950, a body of literature has sought to quantify the adverse effects of misalignment on rotating machinery. Figure 14.8 is simply one of numerous such efforts put forth by highly experienced observers.

Of course, Figure 14.8 is somewhat general and many variables determine its precise accuracy. But it should be used to justify good alignment practices and support the contention that pump misalignment causes bearing overloads and the loss of reliability. Such overloads also create frictional resistance that consumes power. Given the sizeable benefit-to-cost ratio, precision alignment is a key ingredient to optimizing the life cycle cost of process pumps.

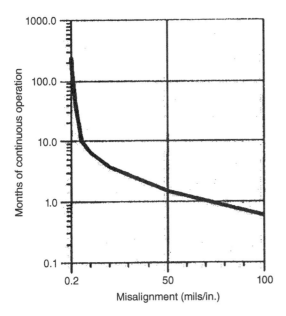

Figure 14.8 Months of continuous operating life vs. misalignment. Source: Modified from Piotrowski [5].

What We Have Learned

- Specify and purchase pump sets with jacking arrangements at driver and pump.
- Grease and oil-mist lubricated drivers require proof that the manufacturer understands both user requirements and relevant technology.
- While not directly considered part of the pump, couplings must be selected with thought given to installation, removal, safety, periodic maintenance, and future performance.
- A size-on-size or 0.0005 in. shaft coupling bore-to-shaft interference fit will minimize shaft fretting and shaft weakening. Careful procedures are needed; they should take into account that this represents a very close clearance.
- Keys should be hand-fitted to shaft keyways.
- Coupling hubs must incorporate removal provisions such as puller holes.
- Reverse dial indicator or laser alignment methods should be used.
- Do not allow shaft misalignments of more than 0.5 mils per inch (0.0005 mm/ mm) of shaft separation (shaft separation is also called DBSE, or distance between shaft ends).

References

1 Bloch, Heinz P., and Shamim, Abdus; *"Oil Mist Lubrication: Practical Applications"*, Fairmont Publishing Company, Lilburn, GA 30047, 1998 (ISBN 0-88173-256-7).
2 Bloch, Heinz P.; *"Practical Lubrication for Industrial Facilities"*, 2nd Edition, Fairmont Publishing, Lilburn, GA 30047, 2009 (ISBN 0-88173-579-5).
3 Bloch, Heinz P., and Budris, Alan; *"Pump User's Handbook"*, 3rd Edition, Fairmont Publishing, Lilburn, GA 30047, 2010 (ISBN 0-88173-627-9).
4 Leibensperger, R. L.; "Look Beyond Catalog Ratings", *Machine Design*, April 3, 1975.
5 Piotrowski, John; *"Shaft Alignment Handbook"*, 3rd Edition, Marcel Dekker, New York, NY, 2006 (ISBN 978-157444-7217).

15

Fits, Dimensions, and Related Misunderstandings

No pump has ever been built (nor will ever be built) for *simultaneously* operating at the extremes of speed, head developed, temperature allowed, limit of dimensional fits, tolerances, pipe-flange induced stress, marginal lubrication, and so forth. Allowing every relevant dimension or parameter to be at their absolute limits *at the same time* will surely cause pumps to fail prematurely and will put people and assets at risk.

Centrifugal process pumps in US oil refineries and other plants typically reach mean-times-between failures (MTBFs) ranging from barely 2 years to as much as 10 years [1]. It can be reasoned that the MTBF of these simple machines is largely influenced by issues or parameters which "someone" either misunderstood or chose to disregard. We have found that many pump failure incidents can be avoided by acting on the information contained in the checklist portion of this chapter.

Low Incremental Cost of Better Pumps

On average, a significantly more reliable pump costs 20% more that its "bare necessity" counterpart. Upgrading during the next repair event is likewise estimated to cost about 20–30% more than a "bare necessity" repair.

Independent estimates from at least two petrochemical corporations hold pump issues responsible for one fire per 1000 pump repair events. With the average pump fire causing several million dollars-worth of damage, achieving fire risk reduction by component upgrading will be worth considerable money. Smart users are, therefore, factoring the imputed value of fire avoidance into their cost justifications.

Pump Wisdom: Essential Centrifugal Pump Knowledge for Operators and Specialists,
Second Edition. Robert X. Perez and Heinz P. Bloch.
© 2022 The American Institute of Chemical Engineers, Inc. Published 2022 by John Wiley & Sons, Inc.

In the late 1990s and early 2000s, the average repair cost of a defective refinery pump was approximately US$11000 – a figure that included burden and overhead [1]. By any measure, implementation of the average improvement item costs less than US$2000 and is worth every penny of it. Our checklist of implementation items lists "things to consider" when pursuing reliability improvement at the point of initial purchase. The same list would apply while carrying out repairs on existing pumps.

A pertinent pump checklist is filled with items, issues, and procedures known to best-practices performers. Many more detailed descriptions can be found in earlier chapters of this text on pump wisdom, others can be found in the reference literature.

Our comprehensive checklist starts with cooling issues, followed by bearing topics, lubrication issues, and then mechanical seal issues. Many of these items or topics overlap; i.e. a cooling topic overlaps with certain lubrication topics and these, in turn, overlap with bearing protector and leakage prevention issues, etc.

Just remember that checklists are not intended to take the place of the rigorous explanations that can be found in narrative texts.

Pump Pedestals and Bearing Housings Should Not Be Water-Cooled

- Do not allow pedestal cooling of centrifugal pumps, regardless of process fluid or pumping temperature. Pedestal cooling is inefficient and corrosion starting from the inside of pedestals has caused massive support failures. If you find a pump with water-cooled pedestals, do the following instead:
  a) Calculate and accommodate thermal growth by appropriate cold alignment offset, per Chapter 14.
  b) Consider using hot alignment verification measurements [2–4].

- Do not allow *jacketed* cooling water application on bearing housings equipped with rolling element bearings:
  a) Note that water surrounding only the bearing outer ring will sometimes cause bearing-internal clearances to vanish, leading to excessive temperatures, lube distress, and premature bearing failure.
  b) Forced-air cooled "finned" bearing housings are acceptable because the bearing outer ring has sufficient clearance in the housing bore so as not to be excessively preloaded by the cooling effect of the air. However, even air-flow cooling may simply not be necessary if superior synthetic oil formulations are used.

- Understand and accept the well-documented fact that, on pump-bearing housings *equipped with rolling element bearings*, it is possible to delete cooling of any kind [5]. For cost and reliability-optimized results,
  a) Simply change over to the correct synthetic lube type and lube formulation. For best results, the lubricant will have to contain proprietary additives.
  b) Use a suitably formulated synthetic lubricant which, at operating temperature, exhibits viscosity characteristics needed for the oil application method utilized in that particular bearing housing.
  c) Recognize that cooling water coils immersed in the oil sump may cool not only the oil but also the air floating above the oil level. The resulting moisture condensation can cause serious oil degradation. This is one of several reasons for avoiding water cooling of pump-bearing housings.
  d) Recognize that *stuffing box cooling* is generally ineffective. If mechanical seals need cooling, investigate alternative seal flush injection and external cooling methods. (In the 1960s, it was proven that with stuffing box cooling the face temperature of mechanical seals is decreased by only about 2 °F, i.e. little more than 1 °C).

- Cooling water may still be needed for effective temperature control in *sleeve bearing applications* (and these are not generally addressed in this text). Again, it should be noted that excessively cold cooling water will often cause moisture condensation.
  a) Even *trace amounts* of water may greatly lower the ability of lubricants to adequately protect the bearing.
  b) On sleeve bearings, close temperature control is far more important than on rolling element bearings.

Summary of Bearing-Related Issues

- Do not use filling-notch bearings in centrifugal pumps. The inevitable axial loading will cause rapid bearing degradation. Replace filling-notch bearings with dimensionally identical deep groove ("Conrad-type") bearings.

- All bearings will deform under load.
  a) On thrust bearings that allow load action in both directions, deformation of the loaded side could result in excessive looseness and hence, skidding, of the unloaded side.
  b) This skidding may result in serious heat generation and thinning-out of the oil film. Metal-to-metal contact will now destroy the bearing.
  c) Always select bearing configurations that limit or preclude skidding. The better bearing manufacturers have application engineering departments

that can advise suitable replacement bearing upgrades. Work with these manufacturers and be prepared to pay a little more for their bearings. It will be well worth it in the long run.

- The API-610 recommended combination of two back-to-back mounted 40° angular contact bearings is not always the best choice for a particular pump application.
  a) Matched sets of well-engineered 40° and 15° angular contact bearings are designed to avoid, or greatly reduce skidding. While not a cure-all, they may thus be more appropriate for a given service application.
  b) On some pumps, sets of 29° or sets of 15° angular contact bearings should be used.

- Keep in mind the possibility of using a 9000-series thrust bearing together with a 7200 or 7300-series angular contact bearing in the thrust location of certain pumps:
  a) Seek factory or application engineering advice and understand best-of-class lubricant application method before proceeding.

- Thrust-bearing axial float, i.e. the total amount of movement possible between thrust bearing outer ring and bearing housing end cap, should not exceed 0.002 in. (0.05 mm). So as to comply with this recommendation, some hand fitting or shimming may be necessary. This will limit what might otherwise become potentially excessive bearing-internal acceleration forces.

- On certain ANSI (American National Standards Institute) pumps, consider replacing old-style double-row angular contact bearings (bearings with one inner and one outer ring) with newer, Series 5300UPG, double-row/double inner ring angular contact bearings.
  a) Series 5300UPG bearings have two brass cages, one outer ring and two inner rings per bearing.
  b) With series 5300UPG bearings, axial clamping of the two inner rings is needed; the advantage is that these newer double-row bearings resist skidding.

- Observe allowable assembly tolerances for rolling element bearings in pumps with single angular contact (Conrad-type), and double-row bearings:
  a) Bore-to-shaft: 0.0002–0.0007″ (0.005–0.018 mm) interference fit
  b) Bearing outside diameter-to-housing fit: 0.0007–0.0015″ (18–37 μm) loose fit.

- Do not use bearings with least-expensive plastic cages in process pumps expected to operate dependably for years:
  a) Some plastic cages tend to be damaged unless highly controlled mounting temperatures are maintained.

b) Plastic cage degradation will not show up in conventional vibration data acquisition and analysis.

c) Superior high-performance plastic cages were developed in the early 2000s. Consider using them if the application engineering group of a competent bearing manufacturer can point to solid experience.

- Be certain to use only precision-ground, matched sets of thrust bearings in either back-to-back thrust or tandem thrust applications:
  a) Matched bearings must be furnished by the same bearing manufacturer.
  b) Verify precision grinding by observing that appropriate alphanumerics have been etched into the back (the wide shoulder) of the outer ring.

- Use radial bearings with C3 clearances in electric motors so as to accommodate thermal growth of the hotter-running bearing inner ring:
  a) Note that modern polyurea greases ("EM" greases) are much preferred for electric motors.

- Investigate column-bearing materials upgrade options on vertical deep-well pumps (earlier shown in Figure 2.3) and compare available high-performance (HP) polymers.
  a) Understand all physical properties and the intended service before picking the right HP polymer bearing for the job.

- Next to oil-jet lubrication, pure (also called "dry-sump") oil mist applied in through-flow fashion per AP-610 (8th and later editions) represents the most effective and technically viable lubrication and bearing protection method used by reliability-focused industry.
  a) Dry-sump (pure) oil mist is also one of the most successful lube application methods for electric motor bearings.
  b) Dry-sump (pure) oil mist is always applied to nonrunning standby pumps and drivers. It protects their bearing housings against the intrusion of airborne moisture and dust.
  c) Taking into account all of the above, investing in oil-mist systems will usually provide paybacks ranging from 8 to 30 months.
  d) Closed oil-mist systems have been in use since the 1960s. They do not allow stray mist to escape to the environment.
  e) Oil mist is virtually maintenance-free.

- Oil-jet system retrofit options are easily engineered for new installations as well as upgrades and conversions.
  a) A simple and economical inductive pump (i.e. a small pump with a free piston as its only moving part) can serve as the source of a continuous stream of pressurized lube oil.

b) Inductive pumps can be used in conjunction with a spin-on oil filter. The resulting clean stream of lubricant can be directed at the bearing rolling elements for optimum effect.

- Realize that oil ring lubrication very rarely represents state-of-art. Oil rings are highly shaft alignment-sensitive and tend not to perform dependably if one or more of the following requirements are not observed:
  a) Unless the shaft system is absolutely horizontal, oil rings tend to "run downhill" and make contact with stationary components.
  b) Ring movement will be erratic and ring edges will undergo abrasive wear. The oil will be seriously contaminated.
  c) The product of shaft diameter (inches) and shaft speed (rpm) should be kept below 8000. Thus, a 3 in. shaft operating at 3600 rpm ($DN = 10\,800$) would not meet the low-risk criteria.
  d) Operation in lubricants that are either too viscous or not viscous enough will not give optimized ring performance and may jeopardize bearing life.
  e) The depth of immersion must be closely controlled, the bore finish must be 16 RMS or better, and the ring eccentricity should not be allowed to exceed 0.002 in. (0.05 mm).

- Flinger spools or flinger discs fastened to pump shafts often perform much more reliably than oil rings (slinger rings):
  a) Consider retrofits using metal flingers discs; realize that cartridge-mounted bearings may be needed.
  b) Some retrofits are made with metal hubs/cores to which elastomeric discs are firmly fused or otherwise attached. Be careful and apply these only within the manufacturer-approved peripheral speed range.

- If the use of oil rings is unavoidable, be aware that a 30° angle between the contact point at the top of a shaft and points of entry into the oil represents the proper depth of immersion. Too much immersion depth will cause rings to slow down, whereas insufficient depth tends to deprive bearings of lubricant.
- Oil rings with circumferentially machined grooves will provide increased oil flow. They still require all of the above risk reduction steps.

Constant Level Lubricators

- Fully consider vulnerabilities of unbalanced constant level lubricators. If you must use constant level lubricators, use only pressure-balanced models.
- Mount constant level lubricators on the correct side of the bearing housing. Observe "up-arrow" provided by manufacturer of constant level lubricators.

Incorrect mounting will lead to disturbances around the air/oil interface in the surge chamber of constant level lubricators. Mounting on the "up-arrow" side of the bearing housing reduces the height difference between uppermost and lowermost oil levels. In other words, it ensures a more limited level variation.

- Recognize that constant level lubricators are maintenance items that will require periodic replacement. The pliable caulking between a transparent bulb and its metal base will degrade over time and small cracks will allow water to be pulled in by capillary action.

Bearing Housing Protector Seals ("Bearing Isolators")

- Consider buying only true state-of-art bearing housing seals to preclude ingress of atmospheric contaminants and egress of lubricating oil. Buy narrow-width products that use only one clamping O-ring and incorporate a sealing O-ring that moves diagonally. Avoid old-style bearing isolators that
  a) use a dynamic O-ring in close proximity to the sharp edges of an O-ring groove (a damaged or nonfunctioning O-ring greatly increases the risk of black oil in the form of O-ring debris)
  b) use a single clamping ring that could cause the rotor to skew or "walk" away from its intended location on the shaft

- Install two large bull's-eye-sight glasses on opposite sides of a pump-bearing housing to view actual operating oil levels.

Motor Lubrication Summary

- Over-greasing of electric motor bearings is responsible for more bearing failures than grease deprivation. Know where the spent grease ends up – hopefully not in the motor windings. Practicing proper regreasing procedures is essential for long bearing life.
- Lifetime-lubricated (sealed) bearings will last only as long as enough grease remains in serviceable condition within the sealed cavity. Whenever the product of bearing bore (mm) multiplied by shaft rotational speed (rpm) exceeds 80 000, reliability-focused plants consider it uneconomical to use lifetime-lubricated bearings in continuously operating industrial machinery.
- Grease replenishing intervals depend on bearing inner ring bore dimension and shaft rotational speed. Reliability-focused user plants consider $dn = 300\,000$ (d = bearing bore, mm; n = shaft rotational speed, rpm) the maximum for grease lubrication of electric motors and other machines in continuous service.

It has been reasoned that beyond this dn-value, grease replenishing intervals become excessively frequent and oil lubrication would be more economical.

- On grease-lubricated couplings, verify that only approved coupling greases are used. Most motor bearing greases will centrifuge apart at high coupling peripheral speeds. Do not use an "EM grease" (electric motor grease) in a gear coupling.
- Conversely, do not allow coupling greases to be used in electric motor bearings. Most motor bearings will fail prematurely unless a premium grade "EM" grease is used.
- The advertised "all-purpose" greases are not suitable for electric motor driver bearings in reliability-focused plants.
- Verify that relubrication and grease replenishment procedures take into account that
  a) Certain grease formulations cannot be mixed with other grease types. Mixing of incompatible greases will typically cause bearing failures within one year.
  b) Attempted relubrication without removing grease drain plugs will often cause the grease cavity to be pressurized.
  c) Over-greasing will cause excessive temperatures. On shielded bearings, cavity pressurization tends to push the shield into contact with rolling elements or bearing cage, causing extreme heat and wear.

Mechanical Seal Issues

- It is generally acknowledged that most pump failure incidents involve mechanical seal distress. While this is true at many facilities, it is also true that a major refinery has documented an average mechanical seal life in excess of 10 years [1]. Using the right selection and installation procedures can markedly improve seal life and reduce pump failure incidents.
- Select mechanical seal types, configurations, materials, balance ratios, pressure–velocity (p–v) values and flush plans certified to represent proven experience in identical services, or under verified-to-be-comparable service and operating conditions. Only these can guarantee to give extended seal life.
- Except for gas seals ("dry seals"), mechanical seals must be operated so as to preclude liquid vaporization between faces. However, using cooling water in a jacketed seal chamber cannot effectively cool the seal environment. External cooling of the flush liquid is far more effective.
- Mechanical seals with quench steam provisions are prone to fail rapidly if quench steam flow rates or pressures are not kept sufficiently low. Installing small diameter fixed orifices will limit excessive steam quench rates.

- Select the optimum seal housing geometry and dimensional envelope to improve seal life. Recognize that slurry pumps generally benefit from steeply tapered seal housing bores. The traditional concentrically bored stuffing box environment does not usually represent the optimum configuration for slurry pumps.
- Avoid inefficient pumping rings (the ones with small cogs machined into the periphery) on dual seals:
  a) Axial pumping screw arrangements are more efficient but, for greater efficiency, some manufacturers use unacceptably close clearances (as little as 0.010 in./0.25 mm) that violate the spirit of 2000-vintage reliability-focused API specification clauses.
  b) Axial pumping screws with close clearances risk making contact with the surrounding sleeve; this could touch off more massive seal failures.
  c) Bi-directional tapered pumping devices for dual seals reduce failure probability; they have open clearances (typically ~0.060 in./1.5 mm) and promote sealing or barrier fluid flow at optimized head and flow rate (H/Q) ratios.

- On hot service pumps, follow approved warm-up procedures. Verify that seal regions are exposed to through-flow of warm-up fluid, i.e. are not dead-ended.
- Understand the difference between conventional mechanical seals (seals where the flexing portion rotates) and stationary seals (where the flexing part is stationary).
  a) Maximum allowable speeds and permissible shaft run-outs are lower for conventional seals and higher for stationary seals.
  b) Seals with springs exposed to the pumpage may be more prone to fail than seals with springs flooded by a more benign liquid or gas (air) environment.

- Flush plans routing liquid from stuffing box back to suction generally require a pressure difference in excess of 25 psi (172 kPa).
- Fluid temperatures in a seal cavity must be low enough to prevent fluid vaporization in seal faces.

Hydraulic Issues

- Determine the suction energy and net positive suction head (NPSH) margin (difference between $NPSH_a$ and $NPSH_r$) for the application. If dealing with a "high" or "very high" suction energy pump (per HI definition), make sure that
  a) the pump is not operating in the suction recirculation region
  b) adequate NPSH margin has been provided
  c) the installation has little or no pipe stress, i.e. good piping practices have been followed

- Ascertain that pumps operating in parallel have closely matched operating points and share the load equally. Examine slopes of performance curves for each. Understand that differences in internal surface roughness may cause seemingly identical pumps to operate at very different flow/pressure points:

    a) Although centrifugal pump life can be greatly curtailed when operating in the low-flow range where impeller-internal flow recirculation is likely to exist, "last resort" help may be of value. A concentric "flow tube" inserted into a spool piece adjacent to the pump suction nozzle will reduce recirculation severity.

    b) On pumps with power inputs over 230 kW, verify that "gap A," the radial distance between impeller disc tip and stationary parts, is in the range of only 0.050–0.060 in. (1.2–1.5 mm). Pump-internal recirculation is thus kept to a minimum. On reduced diameter impellers, this would imply that trimming is done only on vane tips. The impeller covers (shrouds) and discs remain at full diameters.

    c) On pumps with power inputs over 230 kW, ascertain that "gap B," the radial distance between impeller vane tips and cutwater, is somewhere in the range of 6% of the impeller radius. This will reduce vibratory amplitudes occurring at blade passing frequency.

    d) Ensure hydraulically induced shaft deflections in single volute pumps are not excessive. This may require restricting the allowable flow range to an area close to best efficiency point (BEP).

    e) Operate two-stage overhung pumps only at flows within 10% of BEP.

    f) Recognize severe shaft deflection and risk of shaft failure due to reverse bending fatigue when operating far away from BEP.

    g) Check impeller specific speed vs. efficiency at partial flow conditions. Consider installation of more suitable impellers for energy conservation.

    h) Consider installing in the existing pump casing an impeller with different width, or with different impeller vane angle, or different number of vanes, or combinations of these. Observe resulting change trends in performance curves.

    i) Consider changing slope of performance curve by inserting a suitably sized restriction bushing in the pump discharge nozzle.

    j) Review if NPSH gain by cooling the pumpage is feasible and economically justified.

    k) Consider extending the allowable flow range by using an impeller with higher $NPSH_r$. Verify that $NPSH_a$ exceeds the $NPSH_r$ of the new impeller.

    l) Use a ratio $NPSH_a/NPSH_r$ of 3 : 1 or higher for carbamate and similar difficult services. There are many services where several feet or meters of difference between $NPSH_a$ and $NPSH_r$ is not sufficient to prevent caviatation.

m) Be aware of prerotation vortices and their $NPSH_r$-raising effects on mixed flow pumps, i.e. pumps in certain specific speed ranges.

n) Consider use of a vertical column pump or placing pump below grade if $NPSH_a$-gain is needed.

o) Consider inducer-type impellers where lower $NPSH_r$ is needed, but be aware that to the right and left of BEP the new $NPSH_r$ may now actually be higher than before.

p) Mechanical improvement options are often related to specifications, work procedures, and mechanical workforce training. Again, there may be some overlap because every job function in a process plant has potential impact on equipment reliability. Note, also, that a number of items relate to component upgrading. In reliability-focused facilities, every repair event is viewed as an opportunity to consider upgrade options.

Impeller Hydraulics

a) Consider the effects on a performance curve that could result from:
   i) vane underfiling
   ii) vane overfiling
   iii) "volute chipping"

b) Calculate axial thrust values and verify adequacy of thrust disc or calculate balance piston geometry. Modify balance disc or balance piston diameter, as required.

c) Consider opening of existing impeller balance holes if axial thrust must be reduced to extend bearing life. Realize that vendor-supplied balance holes are not always correctly dimensioned and may have to be opened.

d) Review and implement straight-run requirements for suction piping near pump inlet flanges entry. Aim for a straight pipe run of at least five pipe diameters between an elbow and the pump suction nozzle. Consult [1] or Hydraulic Institute guidelines for more precise recommendations.

e) Realize that two elbows in suction piping at 90° to each other tend to create swirling and prerotation. In this case, use 10 pipe diameters of straight run piping between the pump suction nozzle and the next elbow.

f) On top suction pumps, maintain a 10-pipe diameter straight pipe length between suction block valve and pump suction nozzle.

g) Verify that eccentric reducers in suction lines are installed with flat side at top so as to avoid air or vapor pockets. Note the exception in installations with pumped fluid entering suction nozzle from overhead location.

h) Use pumping vanes or suitably dimensioned impeller balance holes to reduce axial load acting on thrust bearings.

i) Oblique trimming of impeller vane exit tips, but retaining equal (Gap "A"-compliant) cover (shroud) and disc diameters, will reduce the severity of vibratory amplitudes at blade passing frequencies and their harmonics.

j) Oblique trimming of impeller vane exit tips, but retaining the full diameter only on the cover (suction side) will make the head-flow curve steeper, while still developing maximum head.

k) "Scalloping" of unsupported regions will improve internal flow profile while reducing risk of cracking and fatigue failure.

Mechanical Improvement or Upgrade Options

- Implement suitable vortex breaker baffles on large vertical sump pumps:

    a) Implement wear ring modifications to reduce severity of rub in the event of contact due to excessive shaft deflection or run-out. Consider using high-performance graphite fiber polyimide resin or perfluoro-alkoxy carbon-filled polymer composites for all wear rings or throat and throttle bushings.

    b) Examine need for occasional measures to cure plate-mode or impeller cover (shroud) vibration. Consider "scalloping," if necessary to avoid impeller vibration other than unbalance-related vibratory action.

    c) Use generous fillet radii (0.2 in., or 5 mm minimum) at shaft shoulders in contact with overhung impellers to avert reverse bending fatigue failures.

- If the calculated shaft deflection exceeds 0.002 in. (0.05 mm) at any of the anticipated flow conditions, the shaft is probably too slender for reliable long-term operation. A superior replacement pump might be considered.

- Verify that shaft slenderness is not excessive. On old API-610/5th Edition pumps, the stabilizing effect of packing may have been lost when converting to mechanical seals. Therefore, throat bushings may have to be replaced by minimum clearance (0.003 in./in., or 0.003 mm/mm shaft diameter) shaft support bushings. These high-performance graphite fiber polyimide resin bushings should be wider than the customary open-clearance throat bushings originally installed.

    a) Verify absence of shaft critical speeds on vertical pumps. Insist on conservative bearing spacing.

    b) Verify acceptability of equipment spacing in pump pits and also ascertain conservatism of sump design. View HI (Hydraulic Institute) guidelines for spacing details.

    c) Consider hollow-shaft motor drivers on vertical pumps and always use suitable reverse rotation prevention assemblies.

d) Beware of exceeding the rule-of-thumb maximum allowable impeller diameter for 3600 rpm overhung pumps: 15 in. (~380 mm).

e) Consider in-between-bearing pump rotors whenever the product of power input and rotational speed (kW times rpm) exceeds 675 000.

f) To survive or to reach long trouble-free operating times, centrifugal pumps must be properly installed. Installation checklists must be used and accountabilities must be defined.

Process Pump Repair Dimensions

All of the following refer to typical refinery pumps. These general guidelines may be used if the manufacturer's more specific instructions are no longer available

1) Radial ball bearing I.D. to shaft fits: 0.0001–0.0007″ (2.5–17 μm) interference

2) Radial ball bearing O.D. to housing fits: 0.0001–0.0015″ (2.5–37 μm) clearance

3) Back-to-back mounted thrust bearing I.D. to shaft fits: 0.0001–0.0005″ (2.5–13 μm) interference

4) Back-to-back mounted thrust bearing O.D. to housing fits: 0.0001–0.0015″ (2.5–37 μm) clearance

5) Shaft shoulders at bearing locations must be square with shaft centerlines within 0.0005″ (13 μm)

6) Shaft shoulder height must be 65–75% of the height of the adjacent bearing inner ring

7) Sleeve to shaft fits are to be kept within 0.001–0.0015″ (25–37 μm) clearance

8) Impeller to shaft fits, on single-stage, overhung pumps, are preferably 0.0000–0.0005″ (0–13 μm) clearance fits

9) Impeller to shaft fits, on multistage pumps, sometimes require interference fits and must be checked against the manufacturer's or in-house reliability professionals' specifications

10) Keys should be located in keyways with 0.0000–0.0001″ (0–2.5 μm) interference

11) In view of (10), above, keys should be hand-fitted to a "snug fit"

12) a) Throat bushing to case fit is to be 0.002–0.003″ (50–75 μm) interference

    b) Throat bushing to shaft fit is to be 0.015–0.020″ (0.4–0.5 mm) clearance

    c) Throat bushing to shaft fits of inline pumps (depending on shaft size) where the throat bushings may act as intermediate bearings, will have clearances ranging from 0.003″ to 0.012″ (0.075–0.27 mm)

13) Weld overlays can be substituted for impeller wear rings. When separate wear rings are used, the wear ring-to-impeller fit should be 0.002–0.003″ (0.05–0.08 mm) interference

14) Impeller wear rings should be secured by either dowelling, set screws that are threaded in the axial direction partly into the impeller and partly into the wear ring, or tack-welded in at least two places

15) Clearance between impeller wear ring and case wear ring should be 0.010–0.012″ plus 0.001″ per inch up to a ring diameter of 12 in. Add 0.0005″ per inch of ring diameter over 12 in.

16) For pumping temperatures of 500 °F (260 °C) and over, add 0.010″ (0.25 mm). Also, whenever galling-prone wear ring materials (such as stainless steel) are used, 0.005″ (0.13 mm) are added to the clearance

17) Impeller wear rings should be replaced when the new clearance reaches twice the original value

18) Case wear rings are not to be bored out larger than 3% of the original diameter

19) Case ring-to-case interference should be 0.002–0.003″ (0.05–0.07 mm)

20) Case rings should be secured (doweled or spot welded) in two or three places

21) Oil deflectors do not take the place of modern bearing protector seals. For best results and if still used, oil deflectors should be mounted with a shaft clearance of 0.002–0.003″ (0.05–0.07 mm)

22) On packed pumps, the packing gland typically has a shaft clearance of 1/32″ (0.8 mm) and a stuffing box bore clearance of 1/64″ (0.4 mm). On packed pumps,
   a) the lantern ring clearance to the shaft is typically 0.015–0.020″ (0.4–0.5 mm)
   b) the lantern ring clearance to the stuffing box bore is typically 0.005–0.010″ (0.13–0.25 mm)

23) Coupling-to-shaft fits are different for pumps below 400 hp (300 kW) driver rating and pumps with drivers of 400 hp and higher. Below 400 hp pumps employ fits metal-to-metal to about 0.0005″ (~13 μm), whereas 400 hp and above models often use interference fits ranging from 0.0005 to 0.002″ (13–50 μm). Taper bore coupling and hydraulic dilation fits should be as defined by either the manufacturer or your in-plant reliability professionals

24) On pumps equipped with mechanical seals:
   a) Seal gland throttle bushing to shaft clearance should typically be 0.018–0.020″ (0.4–0.5 mm), unless the pump is in high temperature (HT) service.
   b) For HT service applications (over 500 °F/260 °C), a somewhat larger clearance may be specified by the manufacturer or your plant reliability professionals.

c) Seal locking collar to shaft dimensions are typically 0.002–0.004″ (0.05–0.1 mm) clearance

25) Heads, suction covers, adapter pieces (if used), and bearing housing to case alignment should have 0.001–0.004″ (0.03–0.10 mm) clearance

What We Have Learned

Checklists or other written reminders are of value to the pump repair person. They merit periodic updating and should be reviewed by personnel representing the three job functions that come into almost daily contact with process pumps: The operators, mechanical/maintenance work force members, and project/technical staff at virtually every facility.

Whenever the specific dimensional guidelines from the pump manufacturer are different from the generalized guidelines found in this text, the manufacturer's guidelines should be used.

If manufacturers' guidelines are no longer available, the experience-based values and dimensions listed here can be implemented with confidence by the process pump user.

References

1 Bloch, Heinz P., and Budris, Alan R.; "*Pump User's Handbook*", 4th Edition, The Fairmont Press, Inc., Lilburn, GA 30047, 2013 (ISBN 0-88173-720-8).

2 Bloch, Heinz P.; "Use Laser-Optics for Machinery Alignment", *Hydrocarbon Processing*, October 1987.

3 Bloch, Heinz P.; "Laser-Optisches Maschinenausrichten", *Antriebstechnik*, Volume 29, Nr. 1, June 1990.

4 Bloch, Heinz P.; "Update Your Shaft Alignment Knowledge", *Pumps & Systems*, December 2003.

5 Bloch, Heinz P.; "*Machinery Reliability Improvement*", 2nd and 3rd Edition, Gulf Publishing Co., Company, Houston, Texas, 1982 (ISBN 0-88415-663-3).

16

Using Failure Statistics and Root Cause Analysis Findings to Guide Reliability Improvement Efforts

It is important to know how well a particular facility's pumps perform compared to those operating in a competitor's facility. There probably is room for improvement if a plant does not measure up to above-average pump failure statistics.

Mean-Time-Between Failures and Repair Cost Calculations

An easy comparison among pump users is feasible. It consists of adding up all process pumps installed at a plant and to then divide by the number of pump repairs per year. For a well-managed and reasonably reliability-focused U.S. refinery with 1200 installed pumps and 156 repair incidents in one year, the mean time between failures (MTBF) would be (1200/156) = 7.7 years.

The refinery would count as a repair incident the replacement of parts – any parts – regardless of cost. In this case, a drain plug worth US$2 or an alloy impeller costing US$5000 would show up the same way on the MTBF statistics. Only the replacement or change of lube oil would not be counted as a repair.

A best-practices plant counts in its total pump repair cost all direct labor, materials, indirect labor and overhead, administration cost and the cost of labor to procure parts. It assigns a value to failure avoidance, even the prorated value of avoiding pump-related fire incidents. Likewise, it assigns a monetary value to a workforce relieved of pump repair burdens and assignable to proactive asset failure avoidance tasks.

Typical published pump repair costs have averaged US$10 287 in 1984 and are estimated to have reached US$17 000 in 2020. After inflation is factored in a repair,

Pump Wisdom: Essential Centrifugal Pump Knowledge for Operators and Specialists,
Second Edition. Robert X. Perez and Heinz P. Bloch.
© 2022 The American Institute of Chemical Engineers, Inc. Published 2022 by John Wiley & Sons, Inc.

an actual cost reduction trend is indicated over this 24-year time span. Predictive maintenance and better pump monitoring may have contributed to reduced failure severity on the typical pump, although the ultimate consequences of some pump failures are grave.

The mean-times-between-failures (installed life before failure) of Table 16.1 have been estimated in 2004. Published data and observations made in the course of performing maintenance effectiveness studies and reliability audits in the late 1990s and early 2000s were used in these estimates.

Seal life statistics were estimated in the early 2000s, Table 16.2. These led to an estimation of reasonable goals, Table 16.3. Note that "target" is less than "best actually achieved."

Many plants achieve the months of *installed* lives indicated in Tables 16.2 and 16.3. Note that the actual *operating* life of a component would thus be about one-half of its installed life. To reach these pump life expectancies, the pump components themselves must be operating at the highest levels of reliability. An

Table 16.1 Pump mean-times-between-failures [1].

ANSI pumps, average, USA	2.5 yr
ANSI/ISO pumps average, Scandinavian P&P plants	3.5 yr
API pumps, average, USA	5.5 yr
API pumps, average, Western Europe	6.1 yr
API pumps, repair-focused refinery, developing country	1.6 yr
API pumps, Caribbean region	3.9 yr
API pumps, best-of-class, U.S. Refinery, California	9.2 yr
All pumps, best-of-class petrochemical plant, USA (Texas)	10.1 yr
All pumps, major petrochemical company, USA (Texas)	7.5 yr

Source: Bloch and Budris [1].

Table 16.2 Suggested refinery seal target MTBFs [1].

Target for seal MTBF in oil refineries	
Excellent	>90 mo
Very good	70/90 mo
Average	70 mo
Fair	62/70 mo
Poor	<62 mo

Source: Bloch and Budris [1].

Table 16.3 Component life targets [1].

	Refineries	Chemical and other plants
Seals		
Excellent	90 mo	55 mo
Average	70 mo	45 mo
Couplings		
All plants	Membrane type	120 mo
	Gear type	>60 mo
Bearings		
All plants	Continuous operation:	60 mo
	Spared operation	120 mo
Pumps		
Based on series system calculation		48 mo

Source: Bloch and Budris [1].

unsuitable seal with a lifetime of one month or less would have a serious negative effect on pump MTBF, as would an under-performing coupling or bearing.

Performing Your Own Projected MTBF Calculations

Simplified calculations will give an indication of the extent to which improving one or two key pump components can improve overall pump MTBF [2].

Say, for example, that there's agreement that the mechanical seal is the pump component with the shortest life, followed by the bearings, coupling, shaft, and sometimes impeller, in that order. The anticipated mean-time-between-failure (*operating* MTBF) of a complete pump assembly can be approximated by summing the individual MTBF rates of the individual components, using the following expression:

$$1/\text{MTBF} = \left[\left(1/L_1\right)^2 + \left(1/L_2\right)^2 + \left(1/L_3\right)^2 + \left(1/L_4\right)^2 \right]^{0.5} \tag{16.1}$$

In a 1980s study, the problem of mechanical seal life was investigated. An assessment was made of probable failure avoidance that would result if shaft deflections could be reduced. It was decided that limiting shaft deflection at the seal face to a maximum of 0.001″ (0.025 mm) probably would increase seal life by 10%. It was thought that increasing seal housing dimensions to accommodate modern seal configurations would more than double seal MTBF.

All components that could be upgraded were examined. The life estimates were collected and then used in MTBF calculations.

In Eq. (16.1), L_1, L_2, L_3, and L_4 represent the life, in years, of the component subject to failure. Using applicable data collected by a large petrochemical company in the 1980s, mean-times-between-failures and estimated values for a reliability-upgraded pump were calculated. The results are presented in Table 16.4. As an example, a standard construction ANSI B73.1 pump with a mechanical seal MTBF of 1.2, bearing MTBF of 3.0, coupling MTBF of 4.0, and shaft MTBF value of 15.0, resulted in a total pump MTBF of 1.07 years (actual *operating* hours). By upgrading the seal and bearings, the estimated achievable pump MTBF (actual *operating* hours) can be improved by 80%, to 1.93 years.

Table 16.4 shows the influence of selectively upgrading either bearings or seals or both on the overall pump MTBF. Choosing a 2.4 year MTBF seal and a six-year MTBF bearing (easily achieved by preventing lube oil contamination via superior bearing housing protector seals) had a major impact on increasing the pump MTBF. Assuming the upgrade cost is reasonable, better seals are the best choice.

Based on year 2002 reports a typical ANSI pump repair costs US$5000. This average cost includes material, parts, labor, and overhead. Assume that the MTBF for a particular pump is 12 months and that it could be extended to 18 months. This would result in a cost avoidance of US$2500/yr – which is greater than the premium one would pay for the reliability-upgraded centrifugal pump.

Reduced power demand would, in many cases, further improve the payback. Selecting advantageous pump hydraulics benefits both pump life and operating efficiency. Audits of two large U.S. plants identified seemingly small pump and pumping system efficiency gains that resulted in power-cost savings of many hundreds of thousands of dollars per year. Thus, the primary advantages of reliability-upgraded process pumps are extended operating life, higher operating efficiency, and lower operating and maintenance costs.

Table 16.4 How selective component upgrading influences MTBF.

ANSI pump upgrade measure	Seal MTBF (yr)	Bearing MTBF (yr)	Coupling MTBF (yr)	Shaft MTBF (yr)	Composite pump MTBF (yr)
None, i.e. "Standard"	1.2	3.0	4.0	15.0	1.07
Seal and bearings	2.4	6.0	4.0	15.0	1.93
Seal housing only	2.4	3.0	4.0	15.0	1.69
Bearing environment	1.2	6.0	4.0	15.0	1.13

Table 16.4 provides a quick means of approximating the annual pump repair frequency based on the total (*installed* life) MTBF. Equation (16.1) and Table 16.4 also can be used to determine potential savings from upgrades and should shape the pump user's strategies.

An experience-based observation assumes that every missed upgrade item reduces pump life by 10% to a new life factor of only 0.9 years. If we miss six such upgrade items (and they are all discussed in the preceding chapters), we will have reduced the anticipated life or MTBF to 0.9E6 = 54% of what it might otherwise have been.

Older Pumps vs. Newer Pumps

After 50 or 60 years of service and many maintenance actions, a large number of "standard" ANSI and ISO-compliant pumps are still operating. When they were designed in the 1950 and 1960s time frame, frequent repairs were accepted. Also, plant maintenance departments were staffed with more personnel than in later decades.

Unless selectively upgraded, a decades-old "standard" pump population will not allow twenty-first century facilities to reach their true reliability and profitability potentials. An older pump will generally fail more often than a newer pump. Likewise, a standard process pump will fail more often than an upgraded process pump.

Buried in Table 16.1 is a plant with over 2000 installed pumps; their average size is close to 30 hp. In 2010, this pump population had an MTBF of slightly over nine years. An informal report placed the 2020 number at 9.6 years. Its owner-operators prided themselves in cultivating effective interaction between the mechanical and process-technical workforce members. The reliability professionals at this plant fully understood that pumps are part of a system and that the system must be correctly designed, installed, and operated if high reliability is to be achieved with consistency. It should also be pointed out that this plant (and others in its peer group) conducted periodic pump reliability reviews.

Reliability Reviews Start Before Purchase

The best time for the first reliability review is before the time of purchase. This subject is given thorough treatment in [3]. Individuals with reliability engineering background and an acute awareness of how and why pumps fail are best equipped to conduct such reviews. Trained reliability professionals should have an involvement in the initial pump selection process. Individually *or as a team* those involved should consider the possible impact of a number of issues, including the ones mentioned earlier. All issues merit close attention and are again summarized for emphasis:

- Keep in mind the potential value of selecting pumps that might cost more initially, but last much longer between repairs. The MTBF of a better pump may be one to four years longer than that of its nonupgraded counterpart.
- Consider that published *average* values of avoided pump failures range from US$2600 to over US$12 000. This does not include lost opportunity costs.
- One pump fire occurs per 1 000 failures. Having fewer pump failures means having fewer destructive pump fires.

Remember that there are several critically important applications where buying pumps on price alone is almost certain to ultimately cause an inordinately larger number of costly failures. Spending time and effort on pre-purchase reliability reviews [3] makes much economic sense, especially when dealing with the following:

- Applications with insufficient NPSH or low NPSH margin ratios [4];
- High-or very-high-suction energy services;
- High specific-speed pumps [5];
- Feed and product pumps without which the plant will not be able to operate;
- High-pressure and high-discharge energy pumps;
- Vertical turbine-style deep-well pumps.

Diligent reviews concentrate on typical problems encountered with centrifugal process pumps; an attempt is made to eliminate these problems before the pumps ever reach the field. Among the most important problems that the reviews seek to avoid are the following:

- Pumps not meeting stated efficiency;
- Lack of dimensional interchangeability;
- Problems with timely delivery because vendor's sales and/or coordination personnel are being reassigned or no longer work there;
- Seal problems and compromises in seal materials, flush plans, flush supplies, etc.;
- Casting voids (repair procedures, maximum allowable pressures, metallurgy, etc.);
- Lube application or bearing problems [6];
- Alignment, lack of registration fit (rabbetted fit), baseplate weaknesses, grout holes too small, base plates without mounting pads, ignorance of the merits of using pregrouted base plates;
- Missing documentation or manuals and drawings shipped too late;
- Pumps that will not perform well when operating away from the best efficiency point, i.e. pumps that risk experiencing internal recirculation [7].

Structured Failure Analysis Strategies Solve Problems

Repeat pump failures are an indication that the root cause of a problem was not found. Alternatively and if the problem cause is known, someone must have decided not to do anything about it. Pursuing a structured failure analysis approach is necessary to solve problems. Guessing or "going by feel" will never do.

Structured analysis means a repeatable approach which can be learned and employed by more than one person [8]. Once an accurate analysis is documented, remedial steps can be agreed upon and can be implemented. Also, whenever it can be established that a pump at location "A" suffers more failures than an identical pump at location "B," we can be sure that an explanation exists. The explanation is found in deviations from best practices in one or more of the following seven cause categories:

- Faulty design
- Material defects
- Fabrication and/or processing (machining) errors
- Assembly or installation defects
- Off-design or unintended service conditions
- Maintenance deficiencies, including neglect/procedures
- Improper operation

Searching for additional cause categories will not add value because anything uncovered will, at best, be a subset of these seven. However, if one systematically concentrates on eliminating five or six of the seven categories in succession, one will arrive at the category where a deviation exists. That will make it possible to concentrate on understanding what led to the deviation.

The pump person must pay close attention to the underappreciated, generally nonglamorous "basics" and do so before opting for the often costly and sometimes unnecessary high-tech solution. Pumps obey the laws of physics, and there is always a cause-and-effect relationship. It follows that even seemingly elusive and generally costly repeat problems can usually be eliminated without spending much money.

An integrated, comprehensive approach to failure analysis starts out by either describing the deviation, or by stating the problem. Next, such an approach encourages, or even mandates, careful observation and definition of failure modes. The approach should employ preexisting or developed-as-you-go checklists and troubleshooting tables [1]. The already existing checklists are supplied by pump manufacturers and can also be found in a very large body of literature.

The "FRETT" Approach to Eradicating Repeat Failures of Pumps

From observation and examination of a failed part one identifies failure agent(s), realizing that there are only four possibilities [8]:

- Force
- Reactive environment
- Time
- Temperature

It is extremely important to accept the basic premise that components will only fail due to one, or perhaps a combination of several, of these four failure agents. We use the acronym "FRETT" to recall these four agents.

Because there are no failure agents beyond these four, the troubleshooter must remain fully focused on these four agents. To reemphasize by an example, a bearing can only fail if it has been subjected to a deviation (or deviations) in allowable force ("*F*"), or has been exposed to a reactive environment ("*RE*"), has been in service beyond its design life ("*T*"), or was subjected to temperatures outside the permissible range ("*T*").

The need for knowledge must not be overlooked. For instance, bearings can fail (overheat) when they are too lightly loaded. They will then skid – a topic that was discussed earlier in this text. But there we go again: Skidding is traceable to an inadequate force ("*F*") and will manifest itself as a temperature excursion ("*T*"). Two of the four agents "FRETT" are at work.

Each failure, and indeed each problem incident, is the effect of a causal event. In other words, for every effect there is a cause; or, there is a reason for every failure. Here's an example:

[Man Injured] – because man fell
  [Man Fell] – because man slipped
    [Man Slipped] – because there was oil on the floor
      [Oil on Floor] – because a gasket leaked

By arriving at the word "gasket," the cause-and-effect chain is focused at the component level. Once we have narrowed issues down to the component level, we know that one or sometimes two troublesome or unexpected or overlooked "FRETT" contributors must now be found. In this case, a gasket leaked. A gasket is clearly a component. So

[Gasket Leaked] – Must be due to: Force? Reactive Environment? Time? Temperature? We must check it out on the basis of data. Without data we would be guessing, and guessing does not lead to repeatable results.
Force: Too much – Why do we rule it in or rule it out?
  Not enough – Why do we rule it in or rule it out?

Reactive environment: Wrong material selected for the medium transported in the pipe? – Why do we rule it in or rule it out?

Time: Was the same gasket left in place for many years? – Why do we rule it in or rule it out?

Temperature: Too high? Too low? Which one of these two (or perhaps why both) might be ruled in or can confidently be ruled out in a particular instance?

The pump person must take a very similar approach with pumps and other machinery. For every effect there is a cause; there is a reason for every failure, and we have to find it:

[Pump is down] – because the shaft broke

[Shaft broke] – our failure mode inventory was consulted; let's assume we found the surface has fretting damage. That is a deviation from the norm.

[Surface damaged] – because the coupling hub was loose. That would explain the fretting damage.

An analyst can now try to get to the root cause by remembering that all pump failure events fit in one or more of the seven cause categories listed above. If the coupling hub was found to be loose, what cause categories are likely and which ones can we eliminate?

- Design Error? Unlikely, since other couplings are designed the same way, and we have verified that they are holding quite well.
- Material Defects? No, since a thorough metallurgical exam checks OK.
- Fabrication Error? No, because the hardness checked OK; dimensional correctness was verified and had been recorded upon installation, three years ago.
- Assembly/Installation Defect? Suppose we have no data and defer it for possible consideration later.
- Off-Design or Unintended Service Conditions? No; we rule it out.
- Maintenance Deficiencies (Neglect/Procedures)? No, since no maintenance (PM) is required on a coupling hub.
- Improper Operation? No, because we have ascertained operator activities were in accordance with our established standards.

At this stage, the analyst would get back to what needs to be investigated further or requires follow-up examination. This might be a good time to start compiling:

a) A checklist of possible assembly errors: From discussions with maintenance personnel, we might conclude that none apply in this instance.
b) A checklist of possible installation errors:
   - *Force*:
     - Could have overstretched hub;
     - Could have had insufficient axial advance on taper (insufficient interference fit).

- *Reactive environment*: None found; normal chemical plant location.
- *Time*: Ascertained that run length was not excessive; the hub failed after just a few weeks of operation.
- *Temperature*: Suppose the coupling was heated to facilitate installation. How was the heat applied? What tells us that the temperature was within limits? The temperature could have been too high (causing overstretch) or too low (not allowing dilation to result in sufficient axial advance).

In both of these examples, the pump failure analyst has to determine in which cause category there is a deviation from the norm, which item needs to be modified, and how this modification must be implemented so as to prevent a repeat failure. Data will be required to support any conclusions. With data one can define the root causes of a problem. Without data one can, at best, determine a probable cause.

Change analysis parallels and supplements the structured, comprehensive approach. Change analysis seeks to identify what is different in the defective item as compared to an identical but unaffected item. The analyst probes into when, where, and why the change occurred. The analyst then outlines a number of remedial action steps and will have to choose the steps that best meet defined objectives. These objectives must achieve highest safety and the analyst may pick from a list that includes lowest life-cycle cost, present value, highest initial quality, meeting a certain industry standard, a deadline, etc.

The objective of aiming for lowest life-cycle cost usually makes considerable sense. Calculating this parameter would include the cost of staffing a pump selection or reliability review with dedicated, knowledgeable individuals. Life-cycle cost analyses must also include the value of downtime avoidance and MTBF extensions, as well as the value of avoided fire and safety incidents.

Recall that fewer pump failures translate into fewer fires and decreased insurance premiums. Failure avoidance creates goodwill and enhances a company's reputation. Also, having to cope with fewer failures frees up personnel whose proactive activities avoid other failures, etc.

Analyzing Pump Failure Data

We will introduce the reader to some basic, yet insightful, pump reliability tracking tools that can be developed with common spreadsheet applications, such as Excel (Figure 16.1). After practicing machinery reliability in the field for many decades, we are familiar with many analysis methods. They range from mathematically complex but not very useful, to simple but providing immediate actionable insights. So while there are numerous ways to present and interpret pump failure data, we believe the methods presented here provide the most information

Figure 16.1 Reliability engineers need data to determine where to focus their improvement efforts. Reliability growth plots and forced ranking tables can turn data into visual tools that help advance the site's reliability improvement efforts.

for the least cost and effort. We hope that readers will also find these methods offered here to be useful and give them a try.

Why We Use Reliability Tools

Useful reliability analysis tools take the available historical failure data and transform them into either visual or concise tabular results. The display will identify reliability problems that require our attention. Here are some of the types of reliability analysis tools we will cover and/or review:

- Pareto failure plots
- Bad actor forced rankings
- Reliability growth plots
- MTBF trends

Let us start by reviewing a simple tool that looks at failures on a sitewide basis. Table 16.5 contains a forced ranking of pump failures for various processing units across a site. By listing the MTBF over the last 12 months, we can quickly identify the potential areas that may need addressing. In the case shown here, the Cat Cracking area seems to be the most problematic of all.

The pump failure data from Table 16.5 can also be converted into a Pareto chart to provide a summary of pump reliability at a glance (see Figure 16.2). (A Pareto chart is a bar graph display, as seen in Figure 16.2, of the frequency that events or measurements appear in a data group of interest.) In our Pareto chart example,

Table 16.5 A hypothetical table of pump failures across a processing facility.

	Number of fluid machinery trains	Number of repairs last year	Total repair cost (US$)	MTBF (mo)
Catalytic cracker	50	34	272 000	17.65
Coker unit	42	21	168 000	24
Crude unit	40	15	120 000	32
Alky unit	35	15	120 000	28
Fractionation unit	40	15	120 000	32
Sulfur unit	25	15	120 000	20
Utilities	42	12	96 000	42
Hydrotreating	32	12	96 000	32
North terminal	30	8	64 000	45
South terminal	32	8	64 000	48

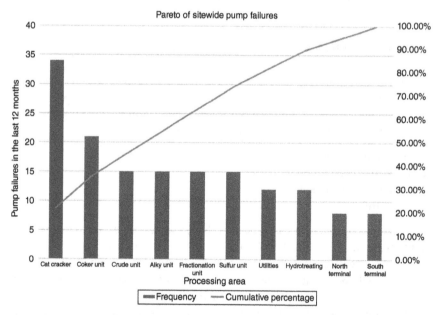

Figure 16.2 Pareto chart of total pump failures over the last 12 months for various processing units. The "cumulative percentage" line helps the reader determine how various groups add to the total failure population. For example, the Cat Cracker and Coker Unit failures represent about 35% of total plantwide pump failures.

pump failure frequencies over the last 12 months for various processing areas are plotted in order of decreasing failure frequency from left to right. Pareto charts are extremely useful for identifying issues that should be addressed first. The reader can quickly see that the Catalytic Cracking area had the most pump repairs over the last 12 months, and that the South Terminal area had the fewest repairs over the same time period. The visual results from this Pareto chart suggest that more study of the Catalytic Cracker Pump failures is warranted.

Now that we know most of the pump failures occurred in the Cat Cracking unit, we can narrow our focus to those pumps. Table 16.6 shows a forced ranking of the pumps with the most failures. In our hypothetical case, Pumps 31-P-09 A&B failed five times in the last 12 months. Assuming that each repair costs about US$10000, we now see that the worst actor cost us about US$50000 in the last 12 months.

You may choose to label the least reliable pumps at your site as "bad actors." Bad actors typically make up 7–10% of the pumps at your site that cost the most to maintain and cause you the most headaches. It makes sense to aggressively address bad actors first.

Cumulative Failure Trends

Management is always interested in knowing if their pump reliability is getting better or worse. A simple means of visualizing historical failure data is by constructing, then analyzing, a special trend called a reliability growth plot, which is a plot of cumulative failures vs. time (see Figure 16.3). This type of graph is constructed by first creating a table of cumulative (total) failures in a populations for consecuative time intervals, then plotting cumulative failures over the time period of interest. For example, let us say that in the first month 20 failures occur in a population, in the second month 25 failures occur, and in the third month 30 failures occur. This would mean the first three points in your reliability growth plot would be the following: Month 1, 20 failures; Month 2, $20 + 25 = 45$ failures; Month 3, $20 + 25 + 30 = 75$ failures, or (1,20), (2,45), and (3,75).

Reliability growth plots allow you to easily see tendencies in the failure data. Figure 16.3 shows three idealized reliability growth plots:

1) A trend where the slope of the cumulative failures is essentially straight, indicating a constant rate of faiiure (shown as "Constant" in Figure 16.3).
2) A trend where the slope of the cumulative failures vs. time sharply increases in July 2016, indicating a decreasing failure rate (shown as "Decreasing" in Figure 16.3).
3) A trend where the slope of the cumulative failures vs. time decreases in July 2016, indicating an increasing failure rate (shown as "Improving" in Figure 16.3).

Table 16.6 A forced ranking of pump failures in the Cat Cracking unit.

Pump	Failures in last 12 months	Total repair costs for the last 12 months (US$)
31-P-09 A&B	5	50 000
31-P-05 A&B	4	40 000
31-P-04 A&B	3	30 000
31-P-08 A&B	3	30 000
31-P-17 A&B	3	30 000
31-P-25 A&B	3	30 000
31-P-02 A&B	2	20 000
31-P-06 A&B	2	20 000
31-P-10 A&B	2	20 000
31-P-11 A&B	2	20 000
31-P-12 A&B	2	20 000
31-P-14 A&B	2	20 000
31-P-16 A&B	2	20 000
31-P-18 A&B	2	20 000
31-P-19 A&B	2	20 000
31-P-22 A&B	2	20 000
31-P-23 A&B	2	20 000
31-P-01 A&B	1	10 000
31-P-03 A&B	1	10 000
31-P-07 A&B	1	10 000
31-P-13 A&B	1	10 000
31-P-20 A&B	1	10 000
31-P-21 A&B	1	10 000
31-P-24 A&B	1	10 000
31-P-15 A&B	0	No repair costs

When readers study a reliablity growth plot, they are able to discern if pump reliability is constant, deteriorating, or improving. Other insights that a reliablilty growth plot provide are when a change in reliabilty occurred, and if the change in reliability was sudden or gradual. In this case, seen in Figure 16.3, we note the deteriorating case indicating that something changed after July 2016. We should look for changes in operating procedures, repair methods, processing rates, etc., to explain changes in pump reliability. Persistent changes in pump reliability may

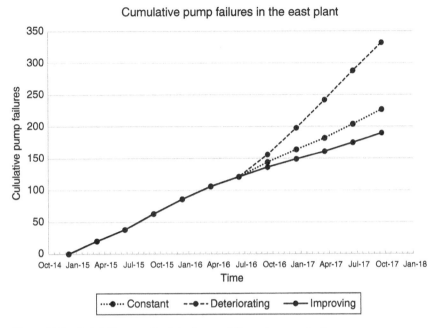

Figure 16.3 Reliability growth plot of pump failures in an operating area.

represent some sort of major change affecting your pump or pumps, while a data blip could simply be mearurement error, or some sporadic factor, such as a plant upset.

Another useful data analysis tool is the MTBF trend plot (Figure 16.4). This type of plot is constructed by plotting a series of MTBF points for a given population over the time-period of interest. The MTBF, which is calculated as shown below, is a simple indicator that provides insight into the mechanical reliability of a single or a group of pumps:

$$MTBF = \frac{\text{Number of pumps} \times \text{Reporting interval}}{\text{Repairs}}$$

For example, if you repair 20 pumps in a 12 months period and you have 50 pumps in the total population, the average MTBF for the year is

$$MTBF = \frac{\text{Number of pumps} \times \text{Reporting interval}}{\text{Repairs}} = \frac{50 \times 12\,\text{months}}{20} = 30\,\text{months}$$

You can use this type of plot to keep an eye on a single or large population of process machines, such as pumps. The MTBF calculated value can be determined

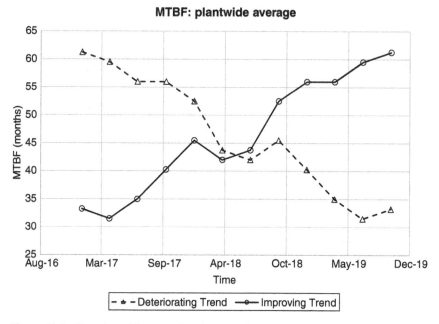

Figure 16.4 Two plantwide mean time between failures (MTBF) trend plots. The solid trend line is an example of an improving MTBF trend, and the dashed line represents a deteriorating MTBF trend.

from the historical data at different reporting intervals then plotted vs. time on a monthly, quarterly, or annual basis. Consider the MTBF trend plot shown in Figure 16.4. By inspection, we can easily compare the MTBF trends of two hypothetical pump populations. We can quickly see that one population (solid line) has a gradually improving MTBF trend and one has a deteriorating MTBF trend (dashed line). The first step to addressing a pump population with a deteriorating MTBF trend (dashed line) would be to drill down into the data of individual process units to see which unit populations are forcing the overall plantwide reliability down. Once the poorly performing process units are identified, you can drill down to the equipment level to identify which pumps are driving the total MTBF trend downward. Identifying the worst performers, also called bad actors, is the first step to improving plantwide reliability.

MTBF: Readers should keep in mind that the MTBF metric is a global metric and only provides limited information about a given machine population. As you drill down to the equipment levels, more advanced analyses, such as Weibull analyses, may be warranted to examine the failure data and to better understand the nature of the failures.

We have briefly covered a few reliability tools that allow you to visualize your equipment's reliability performance. We believe in the usefulness of these methods because we have used them for years with success. In the heat of battle, we need tools that accept easy to collect data and that can be easily analyzed using Excel or similar software applications. Your visual results can then be presented and interpreted by your colleagues and management.

Here is a list of suggestions for developing and maintaining a reliability tracking program:

- Regularly gather centrifugal pump failure data, monthly or quarterly
- Review and "clean" the data to ensure its validity
- Regularly generate reliability plots based on your historical data
- Identify bad actors using the reliability tools presented here. Then, attack bad actors first and report the resulting maintenance cost savings to your management
- Continuously track pump key reliability indicators and look for changes

Addressing Pump "Bad Actors"

Those working around large populations of centrifugal pumps have probably noted that the lion's share of the failures can be attributed to a relatively small proportion of the total population. The Pareto Principle or the 80/20 rule, which posits that 80% of the failures are caused by 20% of the population, describes the reality that every reliability, asset, or maintenance manager must deal with in the real world. The problematic 20% group is made up of individual machines often referred to as "bad actor." History has shown that the best way to improve a site's mean time between repair (MTBR) is to attack "bad actors" first. In time, there will be a gradual improvement in the site's aggregate MTBR value.

A typical "bad actor" sequence of events goes something like this:

1) A new piece of rotating equipment is installed.
2) After the first failure, the mechanics get blamed because everyone assumes the machine was not assembled or installed properly.

> One of the coauthors, Heinz P. Bloch, defined a "bad actor" in his book "*Improving Machinery Reliability*" as follows: "A piece of maintenance-intensive machinery not bad enough to replace with something new or different, but bad enough to drive up maintenance cost, sap maintenance manpower, and cause feelings of resignation or demotivation of personnel."

3) After the second failure, the operator gets blamed because everyone assumes the machine was not started up or maintained properly.

4) After the third failure, the manufacturer gets blamed because everyone assumes the machine was not designed properly to the application.

5) After the fourth failure, it is assumed the failures are related to a deeply rooted systemic issue; so a reliability team is assembled to conduct a root cause analysis.

6) If the team discovers the machine has been misapplied, minor changes are tried initially, such as bearings and seal upgrades. If the resulting reliability is acceptable, you are done.

7) If the acceptable reliability cannot be achieved, then more drastic measures must be taken. These measures may even include replacing the entire machine with one better fitted to the application.

8) If the "bad actor" cannot be replaced, either due to lack of justification or lack of management support capital, then eventually, the operation and reliability departments must learn to live with it. One day, hopefully someone in the organization will take up the challenge to provide a cost-effective solution to obtain a reliable service.

Every plant has its own equipment headaches. Some "bad actors" fly under the radar for years before they are identified. Smaller less critical "bad actors" may hide in the weeds for years, while larger, more critical equipment are quickly identified as problematic and addressed.

We will briefly discuss a methodology the other coauthor, Robert Perez, has used to address "bad actors" as a group. He recommends the following:

1) Force rank equipment (based of MTBR or outage losses) for the area of interest.
2) Select the top 5–10 "bad actors" to attack aggressively.
3) Assign a reliability team to each "bad actor." Teams should consist of the following members:
    a) Process engineer
    b) Reliability professional (leader)
    c) Process operator
    d) Mechanic
    e) Safety professional (as required)
    f) Manager (as required)
    g) OEMs (as required)
4) Compile a detailed database of reliability data for each bad actor
    a) Equipment details
    b) MTBR
    c) Outage information, such as length of outage, economic impact of outage.
    d) Failure plots (reliability growth plots)

5) Whenever a "bad actor" fails, the responsible team members should be informed immediately. It is vital that the repair shop keep the "bad actor" team informed of all failures and any new developments, such as pertinent disassembly observations, assembly issues.
6) The reliability engineer should oversee updating bad actor database files as developments occur.
7) The reliability teams should meet regularly, i.e. monthly or quarterly, to discuss failure details and share improvement ideas.
8) Collectively, all the reliability team members should regularly meet with management to present their progress and recommendations to management.

The roles and responsibilities of "bad actor" team members are shown in Table 16.7. Notice that some members are more involved than others. Some of the members only serve as consultants during the process, while others are simply kept informed. This distribution of responsibilities allows the analysis process to be time efficient. When a machine falls off the "bad actor" list, the team is dissolved, perhaps to join another "bad actor" team.

Management's initial role should be to form teams for each "bad actor." By forming these reliability strike teams, every "bad actor" is made a priority and given the due diligence it deserves. Management needs to also track the program's progress and provide resources required to keep the "bad actor" program moving forward. The hope is that these teams will raise the sitewide awareness of "bad actors" and provide the resources required to ultimately uncover the root causes of failure.

For this approach to be successful, every "bad actor" failure must be scrutinized by the team members so that the root cause is uncovered. Only by knowing a pump failure's root cause can effective design improvements be identified. Detailed failure data are required to chart the best path forward. The strike teams should discuss their findings and then agree on the improvement plan before it is presented to management.

Once recommendations are approved and then implemented, reliability performance should improve along with the site's aggregate MTBR value. Then, when one "bad actor" is deemed to provide acceptable reliability, another bad actor is selected from the updated forced rank asset reliability list. In this manner, improvement of equipment reliability will become evident.

What We Have Learned

Needless to say, any choice we make will have its advantages and disadvantages. When pumps and process pump applications are involved, the most elementary

Table 16.7 Roles and responsibilities matrix for bad actor team.

Task	Reliability Professional	Process Engineer	Process Operator	Mechanic	Safety Professional	OEM	Management Representative
Failure notification	I	I	I	I	I		
Root cause failure analyses	L	P	P	P	C	P	I
Recurring progress meetings	L	P	P	P		I	
Outside lab analyses: metallurgical, oil, field studies, etc.	L			P		C	
Internal research:		L	P				
Internal Lab samples							
Gathering process data							
Tracking progress	L	I	I	I		I	I
MTBF calculation							
Reliability growth plot							
Recommendations	L	P	P	P	C	C	I
Follow-up	L	P	P	P	C	C	I

L = Lead; P = Participant; C = Consulted as needed; I = Informed.

choice requires opting for two out of three broad-brush deliverables: Good, Fast, and Cheap. Take any two, but don't expect to ever obtain all three.

Whenever we are confronted with the two-out-of-three choice, we should remember that for an analysis or repair to be good and fast, it probably will not be cheap. If we want it to be good and cheap, it probably will not be fast. And if we opt to pursue the fast and cheap paths, it probably will not be good. In case we are persuaded to go the fast and cheap route, let us brace ourselves for repeat failures that can cost a small fortune and bring on all kinds of other misery.

Over the decades, we have come to realize that pump failure statistics are rarely very scientific. Still, they are experience-based and should not be disregarded. If your MTBF hovers around average, identify the repeat offenders and subject them to an uncompromising improvement program. In the hydrocarbon processing industry, about 7% of the pump population consumes 60% of the money spent on pump maintenance and repair. Getting at the root causes of failures on these 7% will save much money.

A strategy that involves rational thinking is solidly supported by a minute's worth of looking up vendor documentation. A sound strategy also mandates respect for the simple laws of physics. It is a strategy that results in failure cause identification; it will lead to future failure avoidance and will extend pump MTBF.

It can be said that all successful and cost-effective failure analysis methods represent structured approaches that give focus to an otherwise scattered search for the causes of equipment failures. Structured analysis approaches are repeatable; they aren't hit-or-miss guesses. A successful approach guides the user/analyst through a sequence of steps; it invariably accepts the premise that all problems are ultimately caused by the decisions, actions, inactions, omissions, or commissions of human beings. A successful approach is objective; it seeks explanations but does not tolerate compromises and excuses.

It is fitting, then, to conclude this chapter by pointing to a very simple illustration, Figure 16.5. This illustration tries to convey that many parameters interact to cause repeat failures in pumps. Many of these are classified as hydraulic issues and much work has been done to improve pump hydraulics. However, the majority of what we chose to call *elusive failure causes* is linked to mechanical issues. We have become accustomed to maintenance routines that rarely question the adequacy of a vendor's design. Failure causes have become elusive because we overlook or forget (and even disregard) the laws of physics.

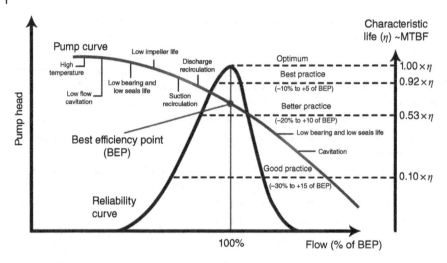

Figure 16.5 Staying near the center of this "Reliability Curve" is a wise course of action. Source: Barringer [9].

References

1 Bloch, Heinz P., and Allan R. Budris; *"Pump User's Handbook"*, 2nd Edition, Fairmont Press, Lilburn, GA 30047, 2006 (ISBN 0-88173-517-5).

2 Bloch, Heinz P., and Don Johnson; "Downtime Prompts Upgrading of Centrifugal Pumps," *Chemical Engineering*, November 25, 1985.

3 Bloch, Heinz P.; *Improving Machinery Reliability*, 3rd Edition, Gulf Publishing Company, Houston, TX, 1998.

4 Karassik, Igor J.; "So, You Are Short On NPSH?", presented at Pacific EnergyAssociation Pump Workshop, Ventura, CA, March 1979.

5 Ingram, J.H.; "Pump Reliability – Where Do You Start", presented at ASME Petroleum Mechanical Engineering Workshop and Conference, Dallas, TX, September 13–15, 1981.

6 Bloch, H.P.; "Optimized Lubrication of Antifriction Bearing for Centrifugal Pumps", ASLE, Paper No. 78-AM-1D-1, presented in Dearborn, MI, April 17, 1978.

7 McQueen, R.; "Minimum Flow Requirements for Centrifugal Pumps", *Pump World*, 6, 2, pp. 10–15, 1980.

8 Bloch, Heinz P., and F. K. Geitner; *"Machinery Failure Analysis and Troubleshooting"*, 3rd Edition, Butterworth-Heinemann Publishing Company, Woburn, Massachusetts, 1997 (ISBN 0-88415-662-1).

9 Barringer, Paul. API Pump Curve Practices from Variability About BEP, Private Weibull Analysis Course (see also Ref. 2, pp. 621).

17

Repair, Replace, or Modify?

When a centrifugal pump fails, there are a number of possible paths forward that a processing facility may consider. They can:

a) Repair the existing pump at an outside shop or inhouse using replacement parts from stores.
b) Replace it with a rebuilt pump (in kind) kept in stores.
c) Replace it with an identical pump kept in stores or purchased from a distributor or manufacturer.
d) Repair it with key upgraded mechanical component(s) to improve mechanical reliability.
e) Perform a hydraulic rerate using the existing pump casing to improve the hydraulic fit.
f) Replace existing pump with a completely different model pump that better fits the service.

It is important that all centrifugal pump maintenance decisions be economically justified, meaning that the benefits derived from the repair, replacement, or modification must be of greater value than the base cost. For maintenance events, such as repairs, the value added by the repair should exceed the cost of the repair. For major modifications, economic criteria, such as the investor rate of return (IRR) and net present value, should be used to objectively evaluate the benefits of the initial investment. The benefits of modifications and replacements are typically increased reliability, an improvement in efficiency, or both. A faithful economic justification requires that the user understand all the benefits derived by the modification along with all the associated modification costs (Figure 17.1).

Pump Wisdom: Essential Centrifugal Pump Knowledge for Operators and Specialists,
Second Edition. Robert X. Perez and Heinz P. Bloch.
© 2022 The American Institute of Chemical Engineers, Inc. Published 2022 by John Wiley & Sons, Inc.

Figure 17.1 A mechanic inspects a centrifugal pump impeller during a repair.

To decide which path forward makes the most economic sense you first need to know:

1) Time since the last repair or the current mean time between failures (MTBF) metric. As we have stated, Best-of Class pumps easily reach MTBF between 6 and 10 years. Pumps with MTBF of less than two years should be considered completely unacceptable. Therefore, pumps providing MTBF performance greater than six years should continue to be repaired as usual and pumps providing MTBF performance less than two years should be scrutinized for improvement opportunities. Pumps with MTBF between two and six years should be reviewed on a case-by-case basis.

2) Root cause(s) of the most recent failure. The pump owners must understand why their pump is failing before an upgrade or replacement pump can be evaluated. The most effective modifications are those that address the root cause of past failures.

3) How the pump is performing and how close to best efficiency point (BEP) is it operating.

4) The cost to restore pump back to manufacturer's specifications.

5) The value of process downtime related to an unplanned pump failure.

6) The cost of an in-kind replacement pump and its delivery time.

7) The cost to keep a replacement pump in stores.

8) The cost of design improvements being considered.

9) The total cost of a centrifugal pump hydraulic rerate being considered.

10) The total cost to replace an existing pump with one that has a better hydraulic fit.

Repair or Replace?

A common question that comes up when pumps fail: Which pump warrants repair and which should simply be replaced? To answer the question, we need the following information concerning the pumps being evaluated [1]:

$m_{salvage}$ = market salvage value
$m_{postrepair}$ = market value of the machine after repair
$r_{value-added}$ = value added to the machine by the repair
r_{cost} = repair cost

Here is the economic test for deciding if a pump should be repaired or replaced: If the value added by the repair ($r_{value\ added}$) is greater than the repair cost minus its salvage value ($m_{repair}-m_{salvage}$), then it makes more sense to repair the pump than replace it. Below is the same statement in equation form:

$$m_{post-repair} - m_{salvage} = r_{value-added} \geq r_{cost}$$

From this relationship, we can conclude that only repairs that produce an added value greater than the cost of repair make economic sense. This statement assumes that the pump in question can be removed from service and repaired without an economic penalty.

Here are a few examples illustrating how the value-added relationship works: Assume, we are considering repairing a 150 hp centrifugal pump:

Pump salvage value = US$ 2000
Postrepair pump value = US$ 20000
Value added = US$ 20000 – US$ 2000 = US$ 18000
Repair cost = US$ 8000

Plugging these values into our value-added formula, we get:

$$m_{post-repair} - m_{salvage} = r_{value-added} = US\$1500 < US\$4000\left(repair\,cost\right)$$

Therefore, a repair is justified in this case since the value added by the repair is greater than the repair cost. Assume we are considering repairing a 5-hp centrifugal pump:

Pump salvage value = US$ 500
Postrepair pump value = US$ 2000
Value added = US$ 2000 – US$ 500 = US$ 1500
Repair cost = US$ 4000

Plugging these values into our value-added formula, we get

$$m_{post-repair} - m_{salvage} = r_{value-added} = US\$1500 < US\$4000\left(repair\,cost\right)$$

In this case, a repair is not justified since the value added by the repair is less than the repair cost.

The reader can see that the lower the replacement cost, the more likely it will make sense to replace a pump than to repair it. Some facilities set a horsepower limit to define when failed pumps are simply replaced, which simplifies the decision-making process for a repair shop.

Repair and Spare Part Philosophies

Once a pump is installed, site management decides its criticality and if it makes sense to repair it or replace it when it fails. Pumps that are fully spared are considered less critical, while pumps that are unspared (nonspared) are considered more critical. There are several cases to consider:

a) For smaller less critical pumps (<5 hp), which are uneconomical to repair, either keep complete pump replacements in stores or order replacement pumps from the manufacturer or distributor when required. The decision to either stock a complete spare pump or order it when required will depend on the expected delivery time.

b) A fully spared pump that can be removed and repaired because the other installed spare is deemed to be reliable and capable of supplying the required flow on its own: In services with medium-to-large size pumps, it makes sense to pull the failed pump and repair it either in the shop or off site.

c) A fully spared pump that can be removed without impacting the process but takes too long to repair: This might apply to troublesome pumps that fail frequently, which may mean that time to repair may approach or exceed the time between failures. In these services, it makes sense to keep a rebuilt or new pump in stores. When needed, the replacement pump can be removed from the warehouse and installed in the field.

d) An unspared pump's downtime must be minimized due to its potential economic impact on the plant: These pumps tend to be more complex and costly, so it makes sense to store a pump bundle or rotating assembly in the warehouse to minimize the repair time in the event of major failures. Spare seals, bearings, and couplings should also be kept in stores for minor repairs.

An ineffective spare parts management program can negatively affect pump reliability at your site. For this reason, spare parts management should be an integral part of every pump reliability program. However, often spare parts management tends to fly below the radar and is often given little thought or consideration. To maintain the effectiveness of the program, a team consisting of a rotating equipment engineer (or professional), shop foreman, warehouse specialist, and a management representative should be involved in all critical spare part decisions and tracking. The team should also meet regularly to discuss: (i) the status of all critical spares, (ii) current spare part stocking levels, (iii) changes in stocking levels, (iv) decisions to dispose of obsolete inventory, (v) addition or deletions of suppliers, (vi) stocking upgraded parts, if available, etc. After some trial and error, an optimal mix of spare parts will be determined for your given centrifugal pump population.

Making the Business Case for Centrifugal Pump Upgrades

Rotating machinery professionals are always looking for ways to improve the reliability or capabilities of their process pumps. However, there is always a limit to the level of improvements that are economically justified. The economical limit is governed by *the law of diminishing returns* (https://www.britannica.com/topic/diminishing-returns), which is an economic law stating that if one input in the production of a commodity is increased, while all other inputs are held fixed, a point will eventually be reached at which additions of the input yield progressively smaller, or diminishing, increases in output.

In a textbook example of the law, a farmer who owns a given acreage of land will find that a certain number of laborers will yield the maximum output per worker. If he should hire more workers, the combination of land and labor will be less efficient because the proportional increase in the overall output would be less than the expansion of the labor force. The output per worker would therefore fall and become less economical.

In centrifugal pump applications, its reliability or process capability are usually considered the commodity of interest, while various design improvement options can be considered inputs to improve reliability or capability to a given economic evaluation. There is a limit to how many economical improvements can be made to any given machine. Eventually, the "nth" improvement will no longer be economically justified. Here are some examples of common centrifugal pump improvements:

- Upgrading mechanical seals and related sealing systems in a pump to improve reliability
- Upgrading pump casing metallurgy to increase useful life
- Installing vibration monitoring equipment to prevent catastrophic failures
- Rerating a pump to allow operation closer to its BEP flow to improve hydraulic stability
- Installing a low flow spillback line in a pump's discharge line to prevent operation below its minimum continuous stable flow (MCSF) to prevent internal recirculation

The payback period method we explain here is a simple means of evaluating the benefits of a potential upgrade to management. The payback period is defined as the initial investment divided by the expected annual revenue realized by the improvement. We determine the payback period of a modification as the total installation cost of the upgrade divided by the annualized benefits, i.e. Payback period = Installation cost/Annualized benefits.

Installation costs can include, but are not limited to, the following items:

- Cost of new components and equipment
- Demolition and installation costs
- Warehouse spares
- Required training and modification/updating of operating procedures

The annualized benefits may include the following:

- Reduction in repair costs
- Reduction in production losses
- Increase in process throughput
- Energy savings

Note that annualized benefits are relative terms. To define an economic benefit, you need to have a future case and a base case. For example, if a pump is failing twice a year and costs US$ 10 000 per repair, then the base maintenance cost is US$ 20 000/yr. If you expect the pump failure rate interval to increase to once every five years, then the future maintenance costs should be US$ 10 000/5 or US$ 2000/yr. In this example, the annualized benefits of the upgrade are expected to be US$ 180 000/yr. The point here is you must always consider the future case and the base case in determining the annualized benefits, i.e. Payback period = Total Cost/DRisk (see DRisk definition below) = Total Cost/(Base Case Risk – Future Risk).

> The annualized risk, or simply risk, is defined here as the sum of all the annualized losses associated with the failure mode being analyzed. Therefore, the annualized risk is equal to the annualized maintenance costs + the annualized process losses + the annualized environmental fines + the annualized demurrage costs + etc. When evaluating the economics of a reliability projects, we need to know the differential between the base risk and the future risk expected from the improvement. This differential is defined as the DRisk, which is the annualized benefit, or revenue, expected to be realized from a reliability improvement project.

For example, assuming a pump failure results in a US$ 20 000 repair, a US$ 50 000 process loss, and a US$ 100 000 demurrage cost, then the losses experienced per seal failure is US$ 170 000. If the pump is failing twice a year, then the annualized base case risk is 2 × US$ 170 000 or US$ 340 000/yr. If we believe an upgraded pump will only fail every five years, then the annualized future risk is expected to be 0.2 × US$ 170 000 or US$ 34 000/yr. Therefore, the annualized DRisk for this project is US$ 340 000/yr minus 34 000/yr or US$ 306 000/yr.

Table 17.1 Likelihood project will be approved, based on the payback period.

Payback period range	Comment
<1yr	Sure to be approved
>1yr but <2yr	Likely to be approved
>2yr but <5yr	Difficult project to sell
>5yr	Not likely to be approved

Table 17.1 tabulates some improvement project payback period ranges along with the likelihood projects that fall in these ranges will be approved by management (based on our experiences).

Payback Time Examples

Example 17.1   Upgrading Mechanical Seals in a Pipeline Pump

Assume there is a multistage (double-ended), pipeline pump that is experiencing a seal failure every six months. Experience has shown that if only the leaking seal is replaced, then the nonleaking seal will normally fail within a few days. For this reason, it is considered good practice to replace both seals when one either of the seal has failed.

RCFAs in the pump's historical record indicate that most of the seal failures are due to abrasives embedding in the softer primary ring and causing a leak. After considering several options, it has been deemed that installing double seals with a dedicated pressurized seal oil system is the most attractive of the upgrade options considered. It is believed that a pressurized sealing

(Continued)

Example 17.1 (Continued)

arrangement will increase the mean time between seal failures from six months to eight years. If the cost of replacing inboard and outboard seals is US$ 40 000, then the annualized reduction in maintenance costs realized from the upgrade is expected to be

$$\frac{US\$40\,000}{0.5\,yr} - \frac{US\$40\,000}{8} = US\$80\,000 - US\$5000 = US\$75\,000\,/\,yr$$

Installation Costs
The upgrade costs include purchasing four seals, two to install and two to put in stores, a pressurized seal oil system, and the cost of installing the required seal auxiliary piping and instrumentation for the new seal oil system. No modifications to the pump shaft or pump casing are required for the seal upgrade. The total project cost of the upgrade is estimated to be US$ 120 000.

What Is the Payback Period for Example 17.1?
Using the payback period formula, i.e. payback period = Total Cost/DRisk = Total Cost/(Base Case Risk − Future Risk), we see that the payback period for this project is US$ 120 000/US$ 75 000 = 1.6 years. Referring to Table 17.1, we can see that this project has a good chance of getting approved. (Notice that this simple economic analysis did not take credit for any pipeline outages or slow-downs caused by unplanned seal replacements. Taking credit for any process-related losses will further improve the payout period and make this upgrade project even more economically attractive.)

Example 17.2   Upgrading Slurry Service Pump Casings

Let us say you own a pair of 500 hp slurry pumps that due to severe casing erosion are currently experiencing one failure every year. Every time the failure occurs, it costs US$ 100 000/yr to pull the pump, replace the casing with a spare casing in stores, and expedite the reassembly and installation. A root cause failure analysis indicates that the current metallurgy and coating combination is unsuited for the application.

Assuming we believe you can improve the MTBF of these pumps from once every year to once every four years, then the annualized benefit of a casing upgrade will be US$ 100 000/yr − US$ 25 000/yr = US$ 75 000/yr. If the cost of the new upgraded casing is US$ 60 000, then the total project cost will be US$ 180 000 (for the main, installed spare, and warehouse spare). We won't include the cost of installing the new casings since future repairs have already been accounted for.

What Is the Payback Period for Example 17.2?
Using the payback period formula, i.e. payback period = Total Cost/DRisk = Total Cost/(Base Case Risk − Future Risk), we find that the payback period for this example will be US\$ 180 000/US\$ 75 000/yr = 2.4 years. According to Table 17.1, a project with a payback greater than two years may be difficult to sell to site management. To increase the chances of getting this upgrade approved, the user may have to find additional upgrade benefits, such as listing any production downtime events caused by unexpected failures or taking a salvage credit for the existing casings.

Remember that the payback period evaluation method does not consider the time value of money as considered in other economic analysis methods. However, most managers tend to find the payback method to be an intuitive and acceptable means of evaluating the economic benefits of smaller projects. Other larger or more complex projects may require more sophisticated economic evaluations, such as determining the IRR or net present value over the life of a project.

Some Final Advice on Upgrades

To be successful, centrifugal pump professionals should (i) use economic thinking when considering all potential upgrade decisions and (ii) always try to justify your projects using clear, well thought-out project scopes that have compelling payback time calculations. Never push improvement projects that are not justified and always listen to your managers' concerns if they question your calculations or assumptions. Don't let your enthusiasm get in the way of your company's best interests.

Centrifugal Pump Hydraulic Rerates

Operations far away from a pump's BEP is a common cause of pump unreliability [2]. There a several reasons why a centrifugal pump may be found to be operating in an off-design condition: The pump may either have originally been misapplied or the process requirements may have changed since its initial specification and installation. Two minimum flows must be satisfied on the pump curve for reliable service: (i) The MCSF, which prevents internal recirculation and (ii) the minimum thermal flow, the point where rapid internal heating begins. The pump manufacturer will usually state the MCSF for their pumps in his documentation. For best overall performance, it is always best to keep the operating point above the hydraulic minimum flow (see Fig. 17.2).

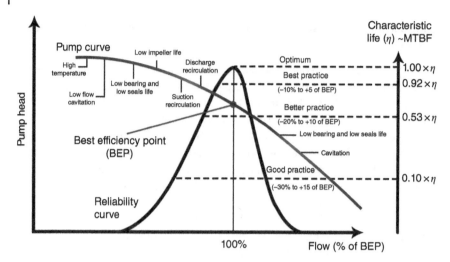

Figure 17.2 The Barringer–Nelson curve shows reliability impact of pump operation away from BEP.

Regardless of the reason of pump misapplication, the effects of a poor hydraulic fit will typically be dismal reliability performance and poor overall system efficiency. Figure 17.2 shows the numerous adverse effects of operating too far away from a pump's BEP sweet spot. In the field, problems related to off-design operations are typically manifested by higher than normal vibration levels and frequent bearing and seal failures.

One solution available to avoid problems caused by an oversized centrifugal pump is to perform a hydraulic rerate on the pump's fluid end. The basic idea of a pump hydraulic rerate is to fit an impeller with a lower flow rating inside the existing oversized pump casing (see Figure 17.3). To accommodate a new impeller, case wear ring adapters are used to allow for smaller diameter sets of casing and impeller wear rings. This would also be an opportune time to consider advanced perfluoralkyl (Vespel CR-6100) stationary wear parts, along with a power end upgrade.

In addition to these modifications, volute throat area(s) must be matched to the new impeller (see Figure 17.4). This is done by cutting out the original volute lips and welding in custom designed volute lips. This simple, straightforward design approach can provide pump users a cost-effective solution to chronic failures caused by hydraulic instabilities.

Hydraulic rerate milestones include the following:

1) Engineering design of volute insert with new throat area
2) Engineering of new impeller with new impeller eye geometry

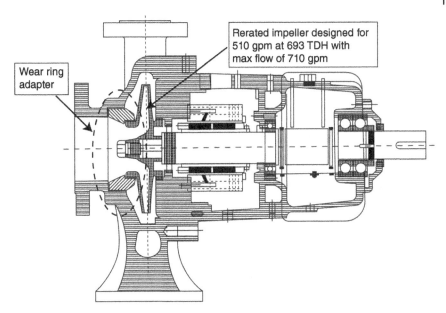

Figure 17.3   Cross section of a rerated pump with new case wear ring adapter and power end upgrade.

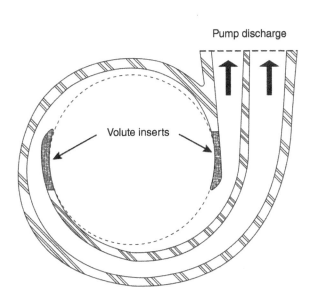

Figure 17.4   Detail of custom-designed volute tongue inserts.

3) Manufacturing and welding in casing volute insert(s)
4) Manufacturing the new impeller and wear rings
5) Manufacture of back pull assembly with mechanical upgrades (if specified)
6) Final pump assembly
7) Performance testing (if required)

Rerate Case Study

Two (main and installed spare) $4 \times 6 \times 13$ single stage, overhung centrifugal process pumps, in lean amine service, coupled to 300 hp/3570 rpm electric motors were installed in a Texas refinery. Table 17.2 shows the combined failure history of both pumps.

This reliability performance equates to an average of 7.8 failures per year and an average MTBF of 0.13 years or 47 days!! Shaft and impeller failures comprised a high percentage of the failures during this time-period. The shaft failures were determined to be related to fatigue. The shaft metallurgy was upgraded during the rash of failures, but the upgrade had little effect on the mean time between failures. After an all-time high of 13 failures in year 6, it was decided to revisit these pumps and perform a detail investigation into the root cause of the shaft failures.

The failure investigation revealed that these pumps (i) had a suction specific speed (N_{ss}) of 16 600, which is well above the recommended design limit of 11 000. To make matters worse, they were operating in a flow range of 31–36% of their BEP (see Figure 17.5). It was obvious that these pumps had no chance to operate reliably.

A pump shop recommended a hydraulic rerate to address both the existing poor hydraulic fit and high N_{ss} design. First and foremost, an available impeller pattern capable of handling the new head/flow requirements and accommodating the

Table 17.2 Combined failure of history of both pumps.

	Number of failures
Year 1	7
Year 2	7
Year 3	7
Year 4	6
Year 5	7
Year 6	13

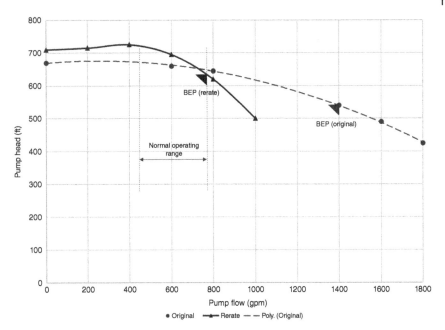

Figure 17.5 Comparison of the original pump performance to the rerated pump performance.

present $NPSH_a$ would need to be located to perform a hydraulic rerate. (Note: Today, if a suitable impeller pattern cannot be found, modern pump facilities can design custom impellers and fabricate them using computerized numerical control [CNC] machines.) In this instance, an impeller pattern with favorable flow, head, and $NPSH_r$ characteristics was located. Table 17.3 lists both the original and the rerated pump impeller design details for comparison.

Figure 17.5 shows the original and rerated pump performance curves, highlighting the dramatic improvement in the hydraulic fit that would be gained by the rerate. It was believed that moving the BEP flow closer to the actual flow requirements (see Figure 17.2) and reducing the suction specific speed below 11 000 with a hydraulic rerate would dramatically improve pump reliability. An expected IRR of 29%, attributable to reliability improvements and energy savings, prompted management's approval without much delay. The rerate shop was told to proceed with the hydraulic rerate.

To preclude the possibility of cavitation, the rerate designers needed to compare the existing NPSH available to the NPSH requirements for the new impeller. Due to the reduction in the required flow, the impeller eye area was decreased from 22 to 13.2 in^2, thereby reducing the suction specific speed from 16 650 to 9636 and increasing the reliable operating range for the modified pump. The reduction in

Table 17.3 Performance comparison of the original and rerated pumps.

	Original design	Rerated design
Design flow (rated)	440–710 gpm	440–710 gpm
Design temperature	170 °F	170 °F
Pump/Impeller size	4″ × 6″ × 13″	3″ × 4″ × 13″
BEP flow	1400 gpm	775 gpm
Volute throat area	5.0 in^2	2.4 in^2
Qd/Q_{BEP}	30–50%	57–92%
N_{ss}	16 650	9636
Q_{min} (estimate)	1126 gpm	320 gpm
Efficiency	55% (at 510 gpm)	67% (at 510 gpm)
BHP	153.4 bhp	130.2 bhp
NPSH$_r$	6.2 ft (at 510 gpm)	12 ft (at 510 gpm)

eye area increased the NPSH requirement from 6.2 to 12 ft. Fortunately, there was more than 12 ft of net positive suction head available in the suction system, which meant the proposed impeller could be used in the rerate.

In addition to the hydraulic rerate, the bearing bracket and stuffing box arrangement were upgraded to conform with the latest API 610 design requirements. The mechanical upgrade also included a shaft with a larger diameter, which improved the rotor's flexibility ratio (L^3/D^4) from 89 to 24. The rotor flexibility ratio improved primarily by enlarging the shaft diameter at the shaft sleeve area through the stuffing box; this minimizes the deflection imposed both the cantilever effect and the radial forces imposed by the impeller.

Once reinstalled and started up, pump vibration levels were clustered near 0.12 ips (3 mm/s). As an added benefit, the operators noticed that pump flows were more stable than before the rerates. As expected, the mechanical reliability of these pumps improved significantly, and no failures were experienced in the first two years of service.

Replacing Existing Pump with an Entirely Different Model Pump

Typically, the costliest way to improve a pump's reliability performance is by completely replacing it with one that is better suited for the application. Initially, site personnel tend to seek improvement with mechanical seal, metallurgy, bearing upgrades, or modifications to operating procedures in their attempts to gain better reliability. Yet, after years of frustration, site management may eventually have to concede that the existing pump or pumps were completely misapplied.

When considering a centrifugal pump replacement, the project group must first find a pump that fits the current hydraulic requirements. If several pump designs are being considered, the pump that requires the least piping modifications is usually favored above the other options. If the selected pump will not fit on the existing foundation, then the cost of the new foundation, or modifying/adapting foundations and/or base plates must be included in the total project cost. Other miscellaneous project costs, such as replacing warehouse spares, training of maintenance and operation personnel, required control modifications, must also be included in the total cost of the project.

Let us go through a pump replacement example to illustrate what is involved in the process:

We will assume we have two 150 hp pumps that are oversized and operate at 25% of the BEP flow. The repair history shows that the pumps have been repaired, on average, once yearly at a cost of US$ 20 000 per repair for the past five years. A review of the pump curve shows that the current hydraulic efficiency is about 27%. The most attractive replacement pump option being considered can provide an efficiency of 67%. The energy savings from the more efficient pump have been calculated to be US$ 46 990/yr (based on an energy cost of US$ 0.097/hp-hr.). We will assume that a new, more reliable pump pair will, on average, have a MTBF of 10 years, which equates to a maintenance cost savings of US$ 20 000– (US$ 20 000/10) or US$ 18 000/yr. These findings mean that the project will attain a total annual revenue stream of US$ 46 990/yr plus US$ 18 000/yr or US$ 64 990/yr. (Note that this total does not include any process losses or additional operational labor costs.)

After researching pump and piping options, the project group has determined that replacing these pumps will cost US$ 500 000, which includes the cost of two new pumps (assuming there is a main and an installed spare), all design costs, recommended spare parts, and all field modifications and installation costs. If we enter these data into an Excel spreadsheet and solve for the internal rate of return and the net present value (NPV), we get the following cash flow table.

By inspection of Table 17.4, we can see that the IRR (14.41%) and NPV (US$ 617 276) for this project are factorable, which is the first hurdle the project will face. Next, management will compare theses economic projections with the other competing projects being considered at the time. Since capital budgets are finite, only the most attractive projects will be funded. The possible outcomes are that the project will either be fully funded, deferred until the next capital budget review, or canceled. Reliability professionals must realize that many reliability improvement projects never see the light of day for reasons outside of their

Table 17.4 Pump replacement economics.

Year		
0	US$(500 000)	Initial Capitol Investment
1	US$ 64 990	
2	US$ 66 940	
3	US$ 68 948	
4	US$ 71 016	
5	US$ 73 147	
6	US$ 75 341	
7	US$ 77 601	
8	US$ 79 930	
9	US$ 82 327	
10	US$ 84 797	
11	US$ 87 341	
12	US$ 89 961	
13	US$ 92 660	
14	US$ 95 440	
15	US$ 98 303	
16	US$ 101 252	
17	US$ 104 290	
18	US$ 107 419	
19	US$ 110 641	
20	US$ 113 960	
IRR	14.41%	
NPV	US$ 617 276.87	
Rate	4%	

control. At the end of the day, remember that only well-conceived reliability improvement projects are ever fully approved and eventually realized. To have a realistic chance of approval, reliability projects need to have excellent economics numbers and key sponsors who will support the project every step of the way. Sponsors can be reliability managers, operations superintendents, plant managers, vice presidents (VPs), etc.

What We Have Learned

- A pump repair only makes economic sense when the value added by the repair exceeds the cost of the repair.
- Smaller, less critical pumps (<5 hp), tend to be uneconomical to repair, so it usually makes sense to either keep complete pump replacements in stores or order replacement pumps from the manufacturer or distributor when required.
- In services with medium-to-large size pumps, it is considered normal to pull the failed pump and repair it either in the shop or off site. This judgment is common whenever the installed spare is deemed to be reliable and capable of supplying the required flow on its own.
- Because larger, unspared pumps are likely to cause economic risks due to unscheduled downtime, it may be appropriate to store a pump bundle or rotating assembly in the warehouse. Lost time due to major repair incidents is thus minimized.
- In cases where pump failure rates are deemed unacceptable or higher than average, reliability improvement considerations may include upgrades to key mechanical components such as bearings, seals, shafts, couplings, and hydraulic optimization while retaining the existing pump casing.
- The ultimate reliability upgrade to consider is the replacement of an existing pump or pump set with completely different model pumps to either improve their hydraulic fit and/or to correct major mechanical design deficiencies.

References

1 Maxham, Jason; *The 50 Percent Rule: Repair or Replace, Revisited*, Published April 25, 2014, in "The Art of Troubleshooting Blog", 2014.
2 Perez, R. X., and Knight, C. Jr.; *Hydraulically Rerating Centrifugal Pumps to Improve Reliability*, Texas A&M Turbomachinery Laboratory, Proceedings of the 15th International Pump Users Symposium, pp 41–49, 1998.

18

Centrifugal Pump Monitoring Strategies

Pump condition monitoring programs can range from the use of portable vibration and temperature measuring devices all the way to a full array of vibration and temperature sensors connected to the distributed control system (DCS) computer system that can store, analyze, and alert nearby operating personnel. The level of pump monitoring used by an organization should be proportional to the associated risk level, i.e. the greater the risk level, the more sophisticated the monitoring system that is justified. Some possible pump monitoring strategies are listed in Table 18.1, along with their advantages and disadvantages.

Keep in mind that as the level of monitoring sophistication increases, the time operators spend around their pumps will likely decrease. While this is probably not a direct relationship, experience has shown that fully automated condition monitoring systems discourage operators from spending time around their process machinery. Manager must find ways to engage their operators to periodically visit the field and closely inspect equipment.

Highly reliability-focused, Best-in-Class users continue to use handheld data collectors because they force operating personnel to visit machine(s) in the field. They realize that only in the field or around processes can operators detect a whiff of steam, hear a squealing belt or coupling, smell H_2S, and so forth. In other words, handheld data collection required operators to leave the comforts of their control centers and walk around their critical pumps. Many machine problems, such as oil leaks, burned paint, piping vibration, unusual sounds, abnormal pressure pulsations can only be detected by trained operators who know their pumps and know what is normal.

Pump Wisdom: Essential Centrifugal Pump Knowledge for Operators and Specialists,
Second Edition. Robert X. Perez and Heinz P. Bloch.
© 2022 The American Institute of Chemical Engineers, Inc. Published 2022 by John Wiley & Sons, Inc.

Table 18.1 Pump monitoring strategies.

Data collection method	Owners	Cost	Advantages	Disadvantages
1 Handheld vibration meter used to spot check pump issues	Operations	Low	Operators must regularly inspect their equipment and gather their own data	Only overall vibration levels are known. Trending relies on manual data collection
2 Vibration data collector is used to periodically collect and store data	Reliability department	Low	Data is archived and can be retrieved and reviewed as needed	Operators may not visit their equipment as often and are less likely to keep track of pump vibration levels
3 1) Strategically placed 4–20 mA vibration transducers tied into local PLC that can shut a pump down on high vibration levels alert and alert local control room of vibration problem 2) Operators must still perform periodic pump inspections to look for developing issues	Maintenance	Low-medium	No trending or spectral information is available	1) Operators may not visit their equipment as often 2) No way to trend data or analyze the spectral content in the vibration data
4 1) Vibration data collector used to periodically gather and store data 2) Strategically placed 4–20 mA vibration transducers tied into local PLC that can shut a pump down on high vibration levels alert and alert local control room of vibration problem. Overall vibration levels can be stored in the process computer via Modbus protocol	Maintenance and reliability	Medium	1) Overall pump vibration levels can be archived on the process computer where they can be trended and analyzed 2) Unexpected vibration excursions between data collection events can be detected	1) Operators may not visit their equipment as often and are less likely to keep track of vibration levels 2) Spectral data is not available

#	Description	Stakeholder	Rating	Benefits	Drawbacks
5	Permanently mounted accels (with analog signal outputs) and temperature sensors with local signal-processing computer connected to control room via Modbus protocol	Maintenance and reliability	High	1) Data is archived and can be retrieved and reviewed as needed 2) Unexpected vibration excursions are captured as they occur	Operators may not visit their equipment as often and are less likely to keep track of vibration levels
6	Permanently mounted accels (with analog signal outputs) and temperature sensors with local signal processing computer connected to control room via Modbus protocol. The data is monitored and analyzed by in-house experts or a third party	1) Maintenance and reliability 2) Third-party analysis company	Very high	1) Data is archived and can be retrieved and reviewed as needed 2) Unexpected vibration excursions are captured as they occur	1) Operators may not visit their equipment as often and are less likely to keep track of vibration levels 2) Vibration analysis expertise resides outside the site or company

Pump Monitoring Recommendations Based on Criticality

Table 18.2 provides a quick means of establishing what pump monitoring methodology should be employed based on the potential risk [1]. To use Table 18.2, start by rating the perceived risk, either low, medium, or high, for each category. The category with the highest level of risk determines the overall pump rating. Then, by determining if the pump is in a local production unit or remotely located, you can select an appropriate monitoring strategy. Keep in mind that this matrix is a starting point. Every organization will have to tweak the descriptors and limits shown here to match their risk tolerance.

How to Use This Matrix

Example 18.1: High Environment Impact (Follow dashed arrows in Table 18.2)

First, we rate the pump criticality in all four categories (economic impact, production impact, etc.). The category netting the highest criticality is the controlling category.

For this example, let us assume that a pump located inside the battery limits of a process unit received the following criticality ratings:

Maintenance impact: Low
Production impact: Low
Environmental impact: **High**
Safety impact: Low

The controlling category is environmental. Therefore, this pump receives "high" as its overall criticality rating. Based on our matrix, it is recommended that this pump be fitted with permanent monitoring instrumentation and shutdowns.

Example 18.2: High Production Impact (Follow dashed arrows in Table 18.3)

The highest risk is a production risk that can potentially lead to a production outage in excess of 24 hours. Suppose the pump is in a remote location. The recommendation here is to install permanent monitoring instrumentation and shutdowns based on a detailed risk analysis and initiate weekly or daily operator inspections (see Table 18.3).

Your site's insurer is another source to consider when deciding the level of condition monitoring your equipment requires to reduce risks to acceptable levels. Based on their knowledge of past failures and associated property damage,

Table 18.2 Pump monitoring recommendations based on criticality.

Criticality rating	Monitoring recommendations	Maintenance impact	Production impact	Environmental impact	Fire risk	Safety impact
High criticality	*Local production unit*: Install hard wired pump sensors tied into a DCS process computer with local PLC shutdowns. Include daily operator inspections *Remotely located*: Same as above	Pump driver >1000 hp	Catastrophic failure leads directly to a production outage lasting >24 hr	Catastrophic failure could lead to a significant environmental release and/or a fine	Catastrophic failure could lead to a major fire	Catastrophic failure could lead to a recordable or lost time injury
Medium criticality	*Local production unit*: Monitor monthly or weekly with data collector and trend. Consider daily operator inspections *Remotely located*: Consider wired loop-powered vibration transmitters with 4–20 mA outputs with automatic PLC shutdowns. Include weekly or daily operator inspections	Pump driver >100 hp but <1000 hp	Catastrophic failure leads directly to significant reduction in production throughput lasting >24 hr	Catastrophic failure could lead to a reportable release but no fine	Catastrophic failure could lead to a significant fire	Catastrophic failure could lead to a first aid

(Continued)

Table 18.2 (Continued)

Criticality rating	Monitoring recommendations	Maintenance impact	Production impact	Environmental impact	Fire risk	Safety impact
Low criticality	*Local production unit:* Monitor quarterly or monthly with data collector and trend *Remotely located:* Monitor monthly with data collector and trend. Consider periodic operator inspections	Pump driver <100 hp	No production consequences	No environmental consequences	Very low risk of fire	No safety consequences

Note: Pump monitoring may include vibration, bearing temperatures, rotor rpm, driver load, and process pressure and flow measurements, which can be either be taken and recorded periodically or monitored continuously depending on the risks involved.

insurers are in a good position to know where your condition monitoring budget is best spent to avoid major losses. If it is determined that a permanent condition monitoring system is justified, then you need to select the optimal condition monitoring system for your application. Now, we will briefly discuss common condition monitoring arrangements.

A Survey of Vibration Sensors

Let us first review the distinct benefits and disadvantages of wireless and wired sensors.

Wireless Sensors

Wireless sensors are becoming more and more common in the world of sensor-based monitoring applications because they are cheap to install, i.e. no cabling is required, and the time spent collecting data by maintenance or reliability personnel is largely eliminated (see Figure 18.1). It should also be noted that wireless sensor networks allow for a greater level of flexibility, because they can be easily adapted to most pump installations. Industrial-grade wireless vibration sensors are outdoor rated and designed to handle environments with liquids, dust, and particulates.

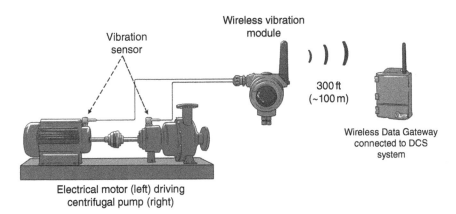

Figure 18.1 Wireless vibration transmitters deliver vibration information over a self-organizing wireless network to a central data acquisition system. Wireless sensors designs are ideal for vibration monitoring applications, especially in hard-to-reach or cost-prohibitive locations. Sensors housings can contain one accelerometer, one accelerometer with embedded temperature, or two accelerometers.

Table 18.3 Pump monitoring recommendations based on criticality.

Criticality rating	Monitoring recommendations	Maintenance impact	Production impact	Environmental impact	Fire risk	Safety impact
High criticality	*Local production unit*: Install hard wired pump sensors tied into a DCS process computer with local PLC shutdowns. Include daily operator inspections. *Remotely located*: Same as above	Pump driver >1000 hp	Catastrophic failure leads directly to a production outage lasting >24 hr	Catastrophic failure could lead to a significant environmental release and/or a fine	Catastrophic failure could lead to a major fire	Catastrophic failure could lead to a recordable or lost time injury
Medium criticality	*Local production unit*: Monitor monthly or weekly with data collector and trend. Consider daily operator inspections. *Remotely located*: Consider wired loop-powered vibration transmitters with 4–20 mA outputs with automatic PLC shutdowns. Include weekly or daily operator inspections	Pump driver >100 hp but <1000 hp	Catastrophic failure leads directly to significant reduction in production throughput lasting >24 hr	Catastrophic failure could lead to a reportable release but no fine	Catastrophic failure could lead to a significant fire	Catastrophic failure could lead to a first aid

Low criticality	*Local production unit*: Monitor quarterly or monthly with data collector and trend *Remotely located*: Monitor monthly with data collector and trend. Consider periodic operator inspections	Pump driver <100hp	No production consequences	No environmental consequences	Very low risk of fire	No safety consequences

Note: Pump monitoring may include vibration, bearing temperatures, rpm, driver load, and process pressure and flow measurements, which can either be taken and recorded periodically or monitored continuously depending on the risks involved.

Despite these benefits, wireless sensors still have some disadvantages. For example, wireless sensors are typically battery powered, so battery life is a concern. However, today's battery powered sensors boast battery operating lives that range from one to six years depending on the analysis requirements. Another disadvantage is that wireless sensor layouts can be limited by the distance constraints between individual transmitters and the data gateway. The speed at which the data are transferred depends on the location of the receiving device relative to the data gateway. By comparison, wired sensors have a much more predictable data transfer time. Another disadvantage of wireless sensors is they can be subject to electrical interference from spurious radio frequency emissions, which may prevent pump condition information from being transmitted at a critical moment. After considering their pros and cons, wireless sensors appear to be better suited to protect pumps at the lower end of the risk ranking spectrum.

Wired Sensors with Dynamic Outputs

Wired accelerometers with analog, i.e. dynamic, outputs are commonly used for industrial machinery monitoring due to their simplicity and reliability. An integral electronics piezoelectric (IEPE) accelerometer incorporates a built-in preamplifier, which transforms the high impedance charge output of a piezoelectric ceramic into a low impedance voltage signal that can be transmitted over longer distances. IEPE sensors are powered by a constant current, typically 2–20 mA, which is supplied by either a signal conditioner or the local data acquisition systems. The output signal from the accelerometer is a voltage signal that is superimposed onto the same line on which constant current is supplied. Coaxial cable is typically used between the accelerometer and signal conditioner or data acquisition system, although most industrial grade accelerometers use shielded, twisted pair cables.

In many situations, wired sensors make up the most reliable monitoring systems, since they are hard wired to the computer that is continuously storing and analyzing the pump vibration and temperature data (see Figure 18.2). It is because of hard wiring that wired vibration sensors are not at risk of the connectivity issues that can plague wireless sensors. However, wired systems require more space and can be more costly to maintain. These factors multiply as more pump sensors are required for a given installation.

Wired Loop-Powered Vibration Transmitters with 4–20 mA Outputs

Standard 4–20 mA output, loop-power vibration transmitters can be used to provide a current output proportional to the overall value of the machine vibration

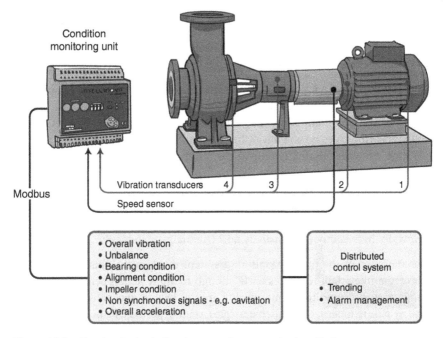

Figure 18.2 How hard-wired vibration transducers can be installed to protect a pump and motor train. Industrial grade transducers with cable protection conduits ensure a mechanically robust installation. A speed sensor may be used for order-tracking failure symptoms. Sensor outputs are sent to the local signal processing unit, which analyzes the dynamic signals to determine if anomalies are present. The local processing unit can also send overall vibration values and alarms, via Modbus protocols, to a nearby control room.

level. When used in conjunction with Modbus compatible process control systems and programmable logic controllers (PLCs), these types of transmitters can provide low-cost machinery protection. Although these do not provide dynamic analog signals, which are required to detect the specific machine faults, they can be used to alarm the machine and indicate when vibration levels are too high. Loop power vibration transmitter makes sense for protecting pumps with a medium-risk level.

We have briefly discussed the pros and cons associated with the three types of condition monitoring systems available on the market: (i) Those based on wireless sensors; (ii) Those based on hard-wired sensors with dynamic outputs; and (iii) Those based on hard-wired sensors with 4–20 mA outputs that are proportional to the overall vibration level. It makes sense to consider wireless and loop-powered sensor for installation on medium- and lower-risk pumps. However, when dealing with highly critical pumps, it is better to entrust your vibration monitoring to field-tested and proven wired sensors with analog (dynamic) outputs than to risk communication problems associated with wireless sensing systems. It is vital that machinery professionals always have access to reliable vibration data to identify pump issues before secondary damage can occur.

The final condition monitoring sensor selection and arrangement chosen for a given pump depends on the answers to the following questions:

1) What type of pump is going to be monitored? Single stage overhung, horizontal multistage, vertical multistage, single stage vertical pump, etc.?
2) What are predominant pump failure modes? Bearings, seals, erosion, etc.
3) What bearing types do you have? Rolling element or hydrodynamic?
4) What is the transmissibility ratio at the bearings, i.e. the ratio of casing vibration to shaft vibration?
5) What is the pump's operating speed range?
6) How will failures of the critical pump components manifest themselves? Vibration, temperature, acoustic energy?
7) How do these components usually fail? Gradually or suddenly?

The designer of the pump monitoring system must ensure that it can detect most of the likely failure modes early enough to allow action before secondary damage occurs. If traditional monitoring approaches are incapable of detecting early key failure modes, then other strategies, such as periodic inspections or component replacements, may be required.

Evaluating Sensor Information

Recall that 4–20 mA signals from loop-powered vibration and temperature transducers are transmitted to the site DCS process computer where they can be archived for later retrieval and review. Alarms will also be communicated to a local control room so that they may be acted upon.

In contrast, sensors generating dynamic signals must be analyzed using a dedicated signal processing computer located near the pump. Complex signal analysis means comparing vibrating levels to present alarm values, generating spectra, evaluating spectral band levels, etc. The length of time that data is stored in the local signal processing computer depends on its memory, the number of data points monitored, and the frequency resolution required. Periodically, the data stored in the signal processing computer can be retrieved for analysis by local reliability personnel.

Alarms and overall measurement information from sensors with dynamic outputs can also be sent to the DCS process computer via Modbus protocol for storage and trending purposes. Here are some examples of how the signal processing computer and site's process computer function together to protect critical process pumps:

If a predetermined vibration or temperature level is exceeded, the alarm event is stored in the local signal processing computer, along with complete set of pump spectra, then an alert is sent to the site's process computer so that the operators know there's a problem. The hope is that the operators will investigate the problem further to determine the present level of risk. The signal processing computer will continue to store alarm events until it runs out of memory. When the memory is full, the newer data will simply overwrite the older data and continue storing new data.

1) An alert can be sent to a local control room if an absolute vibration or temperature level is increasing at a predetermined rate growth. The speed of the increase can aid in determining how much time the site has before they need to act.

2) If a predetermined vibration spectral alarm band (see Figure 18.3) is exceeded, an alert can be sent to the local control room. Operation can then contact reliability professionals to connect to the signal processing computer, inspect all the pump spectra, and determine if there is a problem. Guidelines for spectral bands for centrifugal pump with rolling element bearings are shown in Table 18.4.

Remote monitoring and data analysis may be justified for highly critical pumps or at sites that are unmanned or undermanned. Remote monitoring requires that the local signal processing computer is connected to the Internet via modem, which allows reduced machine data to be remotely accessed and then analyzed by either in-house or third-party condition monitoring experts. Note that most operating sites normally do not allow third-party monitoring hardware to be

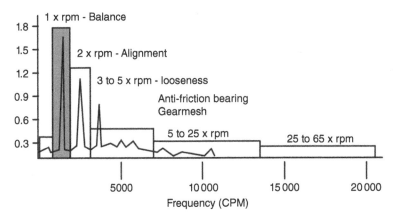

Figure 18.3 Spectral alarm bands. A spectral alarm band is a narrow frequency band within the overall spectrum that is used to monitor changes in a frequency of interest.

Table 18.4 Some examples of how spectral alarm bands can be used to detect mechanical issues commonly modes found in centrifugal pumps with rolling element bearings.

Pump issue	Recommended spectral alarm band setting
Imbalance, misalignment, looseness, or rub	1.5× to 2.5× RPM
Looseness, misalignment, and ball spin frequency	2.5× to 4.5× RPM
First harmonic of bearing fault frequencies (BPFO, BPFI, and BSF)	4.5× to 20.5× RPM
Higher bearing fault frequencies and electrical issues	2.5× to 50× RPM
High frequency acceleration to detect early bearing deterioration, cavitation, or lack of lubrication	1–20 kHz

Note: BPFO stands for ball pass frequency outer race, BPFI stands for ball pass frequency inner race, and BSF stands for ball spin frequency.

connected to the Internet due to security concerns. If an Internet connection is not allowed by the cybersecurity department, pump-monitoring data can be delivered to a third party, via hard drive, thumb drives, security gateways, etc., for analysis.

Make sure your monitoring system has the analysis capabilities required to properly protect you pump installation. Most system will have built-in alarm and trending capabilities but may not have spectral or spectral alarm band capabilities you may need to detect expected failure modes. If you are not sure what analysis capabilities you need, talk to your site's condition monitoring technicians to get a better feel for the types of historical problems they have seen and to which they had to respond.

Vibration Monitoring Analysis Requirements

When selecting a vibration monitoring system, you need to be concerned about its frequency bandwidth capabilities and frequency resolution. This means that the sensors, cabling, the analysis hardware and software must all be integrated properly so that they can detect and report the expected failure mechanisms. For example, a monitoring system on a centrifugal pump with rolling element bearings, operating at 3600 rpm (60 Hz) must be capable of detecting frequencies up to 50 times running speed (3000 Hz) in order to detect bearing failures. Table 18.5 is a good starting point for setting the F_{max} (also called bandwidth) and LOR (lines of resolution) requirements for your system.

Table 18.5 General spectral analysis guidelines for pump and motor assemblies.

Train configuration	Rated speed (RPM)	Bearing type	Orders × RPM	F_{max} (Hz)	Lines of resolution
Pump and motor	900	Rolling element	50	750	800
Pump and motor	1200	Rolling element	50	1000	1600
Pump and motor	1800	Rolling element	50	1500	1600
Pump and motor	3600	Rolling element	50	3000	3200
Pump and motor	900	Sleeve	20	300	800
Pump and motor	1200	Sleeve	20	400	800
Pump and motor	1800	Sleeve	20	600	800
Pump and motor	3600	Sleeve	20	1200	1600

Ensure that the bandwidths and attachment methods of the selected accelerometers will provide the required F_{max} capability for your pump and motor installation [2].
Source: Modified from Graney [2].

Sensors housings are available that contain one accelerometer, one accelerometer with embedded temperature, or two accelerometers. Remember that sensors must be properly located and mounted on the pump to ensure reliable data signals are obtained. Always follow the sensor manufacturer's guidelines on which sensors are right for your application and how to install them properly.

More on Sensor Technology

Before making a major commitment to utilizing permanently mounted pump sensors throughout your plant, the reader should first think about other issues that may be affecting their current reliability performance, such as bad pump selections, specification oversights, and a multitude of opinions and beliefs limiting forward progress. Here are a few ideas to ponder when considering if your valuable reliability resources are best spent on permanent monitoring systems or on addressing known pump reliability issues.

A noted machine sensor authority, Dr. Jan Newby, whom we as pump subject matter experts have known for some time, has contributed valuable corroborating information [3] about remote monitoring systems. He explained that the easiest monitoring systems to install connect directly to the Internet and are accessible via a browser. No server or client software configuration needs to be involved and no dedicated server hardware needs to be sourced. The browser-based software allows easy access from anywhere with an Internet collection.

But the author of Ref. [3] added a precautionary note: The temptation to always configure sensors to collect data very frequently should be discouraged. At Best-in-Class companies (BiCs), most failure modes take months to go from first detection to causing an issue on the asset. Accordingly, collecting data too frequently is simply a waste of time, both in collecting the data and in reviewing it. Excessively frequent collection and review are also a waste of battery life on wireless sensors. Most sensors can be configured with customizable logging (data acquisition) frequencies ranging from minutes to days. In most cases encountered at BiCs, daily or even weekly readings would be more than acceptable. The reasons are rooted in BiCs having invested in above-average asset quality.

In any case, the temptation to always configure sensors to collect data very frequently should be discouraged. Most failure modes take months to go from first detection to causing an issue on the asset; accordingly, too frequent collection is simply a waste of time (both in collecting the data and reviewing it) and also a waste of battery life when using wireless sensors. Most sensors can be configured with customizable logging frequencies; these typically range from minutes to days. In most cases, daily or even weekly readings would be more than acceptable.

On occasion, there are reasons to acquire data more frequently. Examples would include logging data for troubleshooting purposes where process variations may cause faults to appear infrequently, or you need to create a baseline sequence of readings on newly installed or repaired equipment, or where lead time to failure is so short that automatic shut-off of the equipment is required.

But there are some additional questions to ponder and answers to seek before you jump head-first into advocating the widespread use of permanently mounted sensors on pumps. In 2018, both coauthors attended an International Pump Conference. At that conference, questions raised on the exhibit floor occasionally generated a response best characterized as mild annoyance with the ones who dared to "pour rain on our parade." It got us thinking that questions raised by the potential purchaser's reliability professionals would best be answered by respondents on the technology providers' management level. Hopefully, asking these questions will result in factual answers. Sensible answers must take into account the entire sensor-based monitoring and analyzing system, software included. Be sure you know the qualifications of employees or contract personnel that will ultimately use the system. Ask the leading bidder to make a presentation to your employees and managers. Sensible answers coming from competent vendors would (mercifully) steer clear of the often unrealistic projections that reach far into the future.

Not to be neglected is an examination of how BiCs have become BiCs, and there is a tie-in as to why they have not been among the first to embrace remote monitoring technologies as the ultimate solution. BiCs are staffed by competent reliability professionals who have always been strong advocates of predictive maintenance (PdM) methods. Indeed, we view sensor-based technology as the predictive maintenance technology we will see in the future. However, for many years, moves have been underway toward placing progressively greater trust in predictive maintenance. Yet, increased reliance on PdM is not very profitable in situations where failure avoidance by design would have been a smarter strategy. So never lose sight that the overarching aim of this book, the application of "Pump Wisdom," is to attain world reliability across your pump population.

We want you to think of an automobile analogy; it is based on a personal experience with a sporty passenger that came equipped with a 3-l in-line 6-cylinder engine. At 48 000 miles, there were indications of one of its six fuel injectors not working properly. The automobile dealership actually replaced two injectors under the original factory warranty. At 52 500 miles, again two injectors needed exchanging and, although one of the two had been replaced at 48 000 miles, this time the owner was (initially) asked to foot the entire bill. He was told it was "random coincidence." So while it was nice to have advance warnings of impending failure signaled on the dashboard as "service engine soon," it would have been much preferred to incorporate a better injector design in the shiny automobile.

It is no different with your fluid machines and, especially, the widely neglected centrifugal process pumps. Indeed, reliance on PdM may be profitable in hot and/or feed services and certain other situations where equipment failures could create serious safety and environmental problems. In Chapters 3, 4, 5, 14, and 15, we made the point that failure risks greatly escalate whenever several deviations from ideal component dimension, installation practice and pump operating envelope combine, or when lapses in the quality of workmanship occur. While deviations of this type are infrequent, they nevertheless happen. The extent to which sensor technology addresses these concerns remains to be seen.

As just one example, trained professionals understand pump failure events. Slivers of brass indicate oil ring decay and oil rings are a component that works perhaps 97% of the time in pumps. However, when oil rings malfunction, they can cause rapid bearing failures and for sensors to avoid these or to catch them in time is unlikely. How to avoid this kind of event, or lubricant contamination by slivers of abraded oil ring material was explained earlier in this text and in a 2021 text with the title "Optimizing Equipment Lubrication, Oil Mist Technology, and Standstill Protection [4]."

In conclusion, today's remote monitoring sensor technologies have progressed well, and we hope that great successes will be reported as time goes on. Meanwhile, though, it would be prudent for reliability pros to work only with technology providers that can explain how the issues that caused disappointments in earlier decades have now been circumvented or even solved. It is best to do our own digging and find out where a competent provider has installed a full-fledged system that yielded quantifiable benefits. There is also a merit in determining if the reliability managers at such client locations have sought and obtained buy-in from their operators and field maintenance personnel. Once the fact-seeking reliability professionals are fully satisfied, they might ask highly experienced sensor technology providers to assist in compiling reasonably accurate cost justification calculations and case histories. Only then would it be time for both satisfied users and competent turn-key providers to spread the word on the benefits of pump monitoring systems.

What We Have Learned

- Pump condition monitoring programs can range from using portable vibration and temperature measuring devices all the way to a full array of vibration and temperature sensors connected to the DCS computer system. The system can store, analyze, and alert nearby operating personnel.
- Highly reliability-focused, BiC users continue to use handheld data collectors because they force operating personnel to visit machine(s) in the field and stay in touch with the machinery they rely on for reliable operations.
- Selecting the proper pump monitoring methodology involves first evaluating all the types of risk associated with the application and then selecting the methodologies and methods that will mitigate the risks to acceptable levels.
- There are three types of permanent condition monitoring systems available on the market: (i) Those based on wireless sensors; (ii) Those based on hard-wired sensors with dynamic outputs: and (iii) Those based on hard-wired sensors with 4–20 mA outputs that are proportional to the overall vibration level. Consider wireless and loop-powered sensor for installation on medium- and lower-risk pumps and use wired sensors with analog (dynamic) outputs on critical pump.
- When selecting a vibration monitoring system, the responsible machinery professional must be concerned about its frequency bandwidth capabilities and frequency resolution. This means that the sensors, cabling, the analysis hardware, and software must all be integrated properly so that they can detect and report the expected failure mechanisms.

References

1 Perez, R. X., and Conkey, A. P.; "*Is My Machine OK? A Field Guide to Assessing Process Machinery*", Industrial Press Inc., New York, NY, 2012.
2 Graney, Brian P.; Pump Vibration Analysis, *Pumps and Systems Magazine*, 24–28, 2011.
3 Bloch, Heinz P.; "*Fluid Machinery Improvements*", De Gruyter, Berlin, Germany, 2020.
4 Bloch, Heinz P.; "*Optimized Equipment Lubrication, Oil Mist Technology, and Standstill Protection*," 2nd Ed. DeGruyter, Berlin, Germany, 2021.

19

Final Thoughts

In this volume, we have attempted to point out that process pump vendors often merely furnish a barely adequate design. Vendors are left with the impression that users are unwilling to pay for a superior design. Moreover, vendors and pump manufacturers benefit from the sale of replacement parts and are in business to generate income, so they have little motivation to provide designs with enhanced reliability. The generic, good-looking single-stage axially split ANSI pump shown in Figure 19.1 is depicted without enhancements.

We must also not forget that pump manufacturers have right-sized, down-sized, and economized the way they do business. Few (if any) of these organizational realignments benefit the user, and a preponderance of repeat failures attests to it. Some vendors and manufacturers no longer employ process pump experts and diligent craftsmen. User-purchasers may belatedly come to realize that they have become the manufacturer's quality control inspector. Unfortunately, many must experience repetitive failures before they accept this fact. When they eventually learn the hard way about the shortcomings of their pump vendors, they are forced to allocate money to ward off early failures, i.e. infant mortality failures, with suitable predelivery inspections.

Timely and competent up-front action by the owner-purchaser is one of the keys to failure avoidance. This up-front action includes development of detailed specifications for process pumps that address the key components that go into good process pumps. Once a process pump arrives in the field, it must be properly installed and maintained. To be effective, the facility must adopt work processes and procedures that harmonize with best-of-class thinking.

To avoid repeat failures, pump owner-operators must deliberately push certain routine maintenance actions into the superior maintenance category. Superior maintenance efforts will lead to (or are synonymous with) pump reliability upgrading. Its practitioners view every repair event as an opportunity to upgrade.

Pump Wisdom: Essential Centrifugal Pump Knowledge for Operators and Specialists,
Second Edition. Robert X. Perez and Heinz P. Bloch.
© 2022 The American Institute of Chemical Engineers, Inc. Published 2022 by John Wiley & Sons, Inc.

Figure 19.1 There are hundreds of different pump styles and configurations operating in process plants around the world – all are candidates for knowledge-based upgrading.

In essence, the course of wisdom demands that we move away from "business as usual." Before one can apply practical wisdom, one must acquire knowledge and understanding. We hope that this text has helped the reader in this regard.

Index

Pump Wisdom: Essential Centrifugal Pump Knowledge for Operators and Specialists,
Second Edition. Robert X. Perez and Heinz P. Bloch.
© 2022 The American Institute of Chemical Engineers, Inc. Published 2022 by John Wiley & Sons, Inc.